TC 3-21.12

Training Circular
No. 3-21.12

Headquarters
Department of the Army
Washington, DC, 20 July 2012

WEAPONS AND ANTIARMOR COMPANY COLLECTIVE TASK PUBLICATION

Contents

Contents

Figures

Tables

Preface

Purpose

The training circular (TC) is a tool that a commander can use as an aid during training strategy development. The products in this TC are developed to support the company's mission-essential task list (METL) training strategy.

Scope

This TC provides guidance for commanders, staff, leaders, and Soldiers who plan, prepare, execute, and assess training of the weapons and antiarmor company.

Intended Audience

The primary target audience for this TC is the company commander, and other leaders within a weapons and antiarmor company. The secondary audience consists of training developers who develop training support materials for professional military education (PME).

Applicability

This publication applies to the Active Army, Army National Guard (ARNG)/Army National Guard of the United States (ARNGUS), and the United States Army Reserve (USAR) unless otherwise stated.

Feedback

The proponent for this publication is the United States Army Training and Doctrine Command (TRADOC). The preparing agency is the U.S. Army Maneuver Center of Excellence (MCoE). Send comments and recommendations by any means, U.S. mail, e-mail, fax, or telephone, using the format of DA Form 2028, *Recommended Changes to Publications and Blank Forms*. Point of contact information is as follows.

E-mail: BENN.MCoE.DOCTRINE@CONUS.ARMY.MIL
Phone: COM 706-545-7114 or DSN 835-7114
Fax: COM 706-545-8511 or DSN 835-8511
U.S. Mail: Commanding General, MCoE
            Directorate of Training and Doctrine (DOTD)
            Doctrine and Collective Training Division
            ATTN: ATZB-TDD
            Fort Benning, GA 31905-5410

Unless this publication states otherwise, masculine nouns and pronouns may refer to either men or women.

This page intentionally left blank.

Chapter 1

Introduction

As the Infantry's operational environment (OE) evolves, so does the Infantry itself. Infantry units must continually adapt to meet the threat. The Infantry weapons and antiarmor company is a response to meet these changing conditions.

SECTION I – TEXT REFERENCES

1-1. Table 1-1 contains the references used in this chapter.

Table 1-1. Chapter 1 text references

Reference	Subject
ATLDG	Army, G-3/5/7 memorandum, *Army Training and Leader Development Guidance*
ATS	Deputy Chief of Staff, G-3/5/7 memorandum, *Army Training Strategy*
FM 3-21.12	*The Infantry Weapons Company*
FM 3-21.91	*Tactical Employment of Antiarmor Platoons and Companies*
LDS	*The Army Leader Development Strategy for a 21st Century Army*
ADP 3-0	*Unified Land Operations*
FM 7-0	*Training Units and Developing Leaders for Full Spectrum Operations*
FM 6-22	*Army Leadership: Competent, Confident, and Agile*
ATN	Army Training Network link: https://atn.army.mil/index.aspx
FM 1-02	*Operational Terms and Graphics*
ADP 5-0	*The Operations Process*
ADP 6-0	*Mission Command*
AR 350-1	*Army Training and Leader Development*

SECTION II – ARMY APPROACH TO TRAINING

1-2. Before commanders begin planning, preparing, executing and assessing unit training, they must have a clear understanding of the Army's training and leader development strategies, training system, and unit training management (UTM).

ARMY TRAINING STRATEGY

1-3.   The Army goal is to routinely generate trained and ready units for current missions and future contingencies at an operational tempo that is sustainable (Refer to *Army Training and Leader Development Guidance [ATLDG], FY 10-11* for more information.) To accomplish this goal, the Army G-3/5/7 has developed the comprehensive Army training strategy (ATS).

1-4.   The ATS describes the ends, ways, and means required to adapt Army training programs to an era of persistent conflict, to prepare units and leaders to conduct decisive action, and to rebuild strategic depth. The ATS generates cohesive, trained, and ready forces that can dominate at any point on the spectrum of conflict, in any environment, and under all conditions.

1-5.   The ATS has identified ten goals. Each goal has supporting objectives that detail the ATS. Obtaining each goal ensures the Army generates trained and ready units. The goals are—

- Train units for decisive action operations.
- Enable adaption of training.
- Train and sustain Soldier skills.
- Train and sustain Army civilian skills.
- Sustain and improve effectiveness of combat training centers (CTCs).
- Provide training at home station and while deployed.
- Provide training support system live, virtual, constructive, and gaming (LVCG) enablers.
- Increase culture and foreign language competencies.
- Provide supporting and integrating capabilities.
- Resource the Army training strategy.

ARMY LEADERSHIP DEVELOPMENT STRATEGY

1-6.   While the ATS was being developed, the commanding general (CG) of the U.S. Army Training and Doctrine Command (TRADOC) concurrently developed a leader development strategy (LDS). *The Army Leader Development Strategy for a 21st Century Army* discusses how the Army adapts the way it develops leaders. This strategy presents the challenges of the OE, the implications of the OE on leader development, and the mission, framework, characteristics, and imperatives of and how to implement the strategy. The LDS describes eight specific imperatives to guide the policy and actions necessary to produce the future leaders the Army needs.

1-7.   The LDS is part of a campaign of learning. It seeks to be as adaptive and innovative as the leaders it must develop. The LDS is grounded in Army leadership doctrine and seeks to deliver the leader qualities described in both Army doctrine and capstone concepts. (Refer to FM 6-22 and *The Army Leader Development Strategy for a 21st Century Army* for more information.) ADP 3-0 describes how the Army seizes, retains, and exploits the initiative to gain and maintain a position of relative advantage in sustained land operations through simultaneous offensive, defensive, and stability operations in order to prevent or deter conflict, prevail in war, and creates the conditions for favorable conflict resolution.

ARMY TRAINING SYSTEM

1-8.   The Army Training System prepares Soldiers, Army civilians, organizations, and their leaders to conduct decisive action. The training system is built upon a foundation of disciplined, educated, and professional Soldiers, civilians, and leaders, adhering to principles that provide guidance.

Principles of Unit Training

1-9.   To maintain a professional baseline the Army has developed 11 training principles that govern Army training. (Refer to FM 7-0 for more information.) The principles provide a broad but basic foundation to guide how commanders and other leaders plan, prepare, execute, and assess effective training. The 11 principles of training are:

- **Commanders are responsible for training units.** The unit commander is the unit's primary training manager and trainer. Commanders hold their subordinate leaders responsible for training their respective organizations. This responsibility applies to all units in both the operational Army and the generating force.
- **Noncommissioned officers train individuals, crews, and small teams.** Noncommissioned officers (NCOs) are the primary trainers of enlisted Soldiers, crews, and small teams. Officers and NCOs have a special training relationship; their training responsibilities complement each other. This relationship spans all echelons and types of organizations. Noncommissioned officers are usually an organization's most experienced trainers.
- **Train to standard.** Army training is performed to standard. Leaders prescribe tasks with their associated standards that ensure their organization is capable of accomplishing its doctrinal or assigned mission. A standard is the minimum proficiency required to accomplish a task under a set of conditions.

- **Train as you will fight.** "Fight" includes lethal and nonlethal skills in decisive action. "Train as you will fight" means training under the conditions of expected, anticipated, or plausible OEs.
- **Train while operating.** Training continues when a unit is engaged in operations. Combat builds experience but not necessarily effectiveness. To adapt to constantly changing situations, units continue to train even in the midst of campaigns.
- **Train fundamentals first.** Fundamentals include warrior tasks and battle drills as well as METL tasks. Company-level units establish the foundation. They focus their training on individual and small-unit skills. These tasks typically cover basic soldiering, drills, marksmanship, fitness, and military occupational specialty proficiency.
- **Train to develop operational adaptability.** Although planning is critical to successful training, circumstances may cause plans to change. Leaders prepare for personnel turbulence and equipment shortages even though the Army forces generation (ARFORGEN) system tries to ensure personnel and equipment objectives are met before training begins.
- **Understand the operational environment.** Commanders understand the OE and how it affects training. They replicate operational conditions, including anticipated variability, in training. The essence of the principle is to replicate conditions of the OE as part of training to standard.
- **Train to sustain.** Units must be able to operate continuously while deployed. Essential for continuous operations, sustainment is an integral part of training.
- **Train to maintain.** Commanders allocate time for units to maintain themselves and their equipment to standard during training events. This time includes scheduled and routine equipment maintenance periods and assembly area operations. Leaders train their subordinates to appreciate the importance of maintaining their equipment. Organizations tend to perform maintenance during operations to the standards they practice in training.
- **Conduct multiechelon and concurrent training.** Multiechelon training is a technique that allows for the simultaneous training of more than one echelon on different or complementary tasks. It is the most efficient way to train, especially with limited resources. It requires synchronized planning and coordination by commanders and other leaders at each affected echelon.

Principles of Leader Development

1-10. Leader development is deliberate, continuous, and progressive, spanning a leader's entire career. Leader development comprises training and education gained in schools, the learning and experiences gained while assigned to organizations, and the individual's own self-development.

1-11. Every Army leader is responsible for the professional development of subordinate leaders, military and civilian, and for building and sustaining the leader characteristics and skills. (Refer to FM 6-22 for more information.) Company commanders are responsible for leader development of subordinates and are every leader's top priority. Effective training and education build good leaders, and good leaders develop and execute effective training and education in schools and units. The experience gained during assignments puts the training and education into practice and provides the skills and knowledge leaders need to be versatile, adaptable, well-rounded, competent professionals. The Army's principles of leader development are:

- **Lead by example.** Leaders are role models. To demonstrate good leadership is to teach good leadership. Everything a leader does and says is scrutinized, analyzed, and often imitated. The example set by commanders influences the thoughts and attitudes of their subordinates, their families, and their peers. A good example positively influences the development of subordinates.

- **Take responsibility for developing subordinate leaders.** Commanders take responsibility for developing subordinate leaders. They directly observe, assess, and provide honest informal and formal feedback to subordinates. They discuss ways to sustain and improve leader skills, abilities, behaviors, and knowledge with subordinate leaders as often as needed, and ensure subordinates undergo experiences that prepare them for success, improve their adaptability, and prepare them for future responsibilities. They ensure their subordinates attend professional military education at the right time in their careers and functional training to make them effective leaders in their units of assignment.

- **Create a learning environment for subordinate leaders.** Leaders learn in an environment conducive to growth. Growth occurs best in environments that provide subordinates with opportunities to overcome obstacles and make difficult decisions. Commanders encourage their subordinates to seek challenging assignments, and commanders underwrite subordinates' honest mistakes. Learning comes from both successes and failures. Leaders must feel comfortable taking risks and trying new approaches to training. An environment that allows subordinate leaders to make honest—as opposed to repeated or careless—mistakes without prejudice is essential to leader development.

● **Train leaders in the art and science of mission command.** Commanders approach mission command training from two perspectives. First, they train themselves and their subordinates on how to conduct operations using mission command. (Refer to ADP 3-0 and ADP 6-0 for more information.) Second, they follow the principles of mission command in UTM. Specifically, they tell their subordinates the purpose for training and the end state they expect from it, but they leave the determination of how to achieve the end state to the subordinate. As appropriate, they provide guidance requested by the subordinate leader. Employing mission command in training follows the principle of "train as you will fight." Using mission command principles improves not only mission command skills, but it also encourages risk-taking, initiative, and creativity.

● **Train to develop adaptive leaders.** The Army continues to succeed under the most challenging conditions because Soldiers and Army civilians adapt to unexpected situations. Operational adaptability begins in the schools and is then put into practice during tough, realistic training situations—well before leaders are engaged in decisive action. Knowing that changes occur, effective commanders plan for it and develop potential contingency plans to mitigate the effects of change. Effective commanders also look for indicators that change is about to occur so they can ease the transition effects. Placing subordinate leaders into changing, unfamiliar, and uncomfortable situations in training helps foster operational adaptability. The lessons they learn help develop intuition, confidence, and the ability to think on their feet. The Army trains leaders for their next position before they assume it. Cross-training provides unit depth and flexibility and builds leader confidence.

● **Train leaders to think critically and creatively.** The Army develops leaders able to solve difficult, complex problems. Leaders should be able to recognize the issue, quickly ask the right questions, consider a variety of alternative solutions, and develop effective solutions. They should be comfortable making decisions with minimal information. (Refer to ADP 5-0 for more information.)

● **Train leaders to know their subordinates and their families.** Every commander should know his subordinates at least two levels down—their strengths, weakness, and capabilities. An effective leader maximizes a subordinate's strengths and helps him overcome weaknesses. Similarly, an effective leader provides advice, counsel, and support as subordinate leaders develop their own subordinates. Family well-being is essential to unit and individual readiness. The Army trains leaders to know and help not only the subordinates, but also their families. Training ensures subordinate leaders recognize the importance of families and are adept at helping individuals solve family issues and sustain sound relationships.

UNIT TRAINING MANAGEMENT

1-12.   Unit training management is the process used by Army leaders to identify training requirements and subsequently plan, prepare, execute, and assess training. UTM provides a systematic way of managing time and resources and of meeting training objectives through purposeful training activities.

1-13.   The commander's role in training focuses on determining which tasks the unit trains based on the mission. Unit leaders understand the unit's mission and the expected operational conditions to replicate in training. From this, the commander identifies collective tasks to train and the associated risks of not training other collective tasks to proficiency. The conditions are described in the higher unit's training and leader development guidance, or are likely to be encountered in a mission. The commander visualizes the unit's required state of readiness for the mission and the training necessary to achieve METL proficiency, given the commander's assessment of current task proficiency. The commander describes the training plan in training and leader development guidance or operation orders and directs its execution. By participating in and overseeing training and listening to feedback from subordinates, commanders assess the unit's METL proficiency and whether the training being conducted contributes to mission readiness

1-14.   Unit training management is the practical application of the training doctrine. The UTM information is contained in FM 7-0.

1-15.   FM 7-0 and UTM are posted within the Army Training Network (ATN) . The ATN is an Internet website that provides best practices, examples, tools, and lessons learned. It also provides a wealth of other training resources to include the latest training news, information, products and links to other Army training resources.

1-16.   These references are linked and designed to be used in concert as a digital resource. FM 7-0 provides the intellectual framework of what Army training is, while UTM provides the practical how-to of planning, preparing, executing, and assessing training in detail. The ATN, as the digital portal to both documents, additionally provides a wealth of other training resources to include the latest training news, information, products and links to other Army training resources.

ARMY FORCE GENERATION

1-17. Army force generation (ARFORGEN) is a process that progressively builds unit readiness over time during predictable periods of availability to provide trained, ready, and cohesive units prepared for operational deployments. (Refer to FM 7-0 for more information.)

1-18. ARFORGEN drives UTM within the Army. (Refer to FM 7-0 for more information.) Unit training management is the process used by Army leaders to identify training requirements and subsequently plan, prepare, execute, and assess training. Army UTM provides a systematic way of managing time and resources and of meeting training objectives through purposeful training activities.

1-19. The Army prepares and provides campaign capable, expeditionary forces through ARFORGEN, which applies to Regular Army (RA) and Reserve Component (RC) units (Army National Guard and U.S. Army Reserve).

1-20. ARFORGEN takes each unit through a three-phased readiness cycle (known as pools): reset, train/ready, and available. The reset, train/ready, and available force pools provide the framework for the structured progression of increased readiness in ARFORGEN. (Refer to AR 350-1 for more information.) The force pools are defined as follows:

- **Reset force pool.** Units enter the reset force pool when they redeploy from long-term operations or complete their window for availability in the available force pool. The RA units remain in the reset force pool for at least 6 months, and RC units remain in the reset force pool for at least 12 months. Units in the reset force pool have no readiness expectations.

- **Train/ready force pool.** A unit enters the train/ready force pool following the reset force pool. The train/ready force pool is not of fixed duration. Units in the train/ready force pool increase training readiness and capabilities as quickly as possible, given the resource availability. Units may receive a mission to deploy during the train/ready force pool.

- **Available force pool.** Units focus on deployment and training to sustain METL fundamentals and correct any operational deficiencies. Units in the available force pool are at the highest state of training and readiness capability and are ready to deploy when directed. The available force pool window for availability is one year.

1-21. Units move from the available force pool to the reset force pool following a deployment or the end of their designated window of availability.

SECTION III – OTHER TRAINING CONSIDERATIONS

1-22. In addition to understanding the ATS and UTM, commanders should also consider—

- Operational environment.
- Decisive action operations.
- Training products.

OPERATIONAL ENVIRONMENT

1-23. An OE is a composite of the conditions, circumstances, and influences that affect the employment of military forces and bear on the decisions of the unit commander. The complex nature of the OE requires commanders to simultaneously combine offensive, defensive, and stability or civil support tasks to accomplish missions domestically and abroad. (Refer to ADP 3-0 for more information.)

OPERATIONAL VARIABLES

1-24. Company commanders and other leaders analyze and describe the OE in terms of operational variables. Commanders continually monitor their operational environment at the tactical level consistent with mission variables. They apply the military aspects of terrain as a means of protecting the force. Commanders also find it useful to use the operational environment variables as a method to analyze information. Information is used to clarify the evolving operational, tactical, and criminal threat picture for commanders through pattern analysis and the information assessment process (IAP). The company must be trained, adaptable, and ready to operate effectively on short notice. It must possess a wide range of skills, proficiencies, and capabilities to function effectively in any OE.

1-25. These operational variables are easily remembered using political, military, economic, social, information, infrastructure, physical environment, and time (PMESII-PT). (Refer to ADP 3-0 for more information.)

MISSION VARIABLES

1-26. Operational variables may be too broad for tactical planning. Upon receipt of a warning order or mission, commanders and leaders should narrow their focus to six mission variables. Mission variables are those aspects of the OE that directly affect a mission. They outline the situation as it applies a specific Army unit. Mission variables are mission, enemy, terrain and weather, troops and support available, time available and civil considerations (METT-TC). (Refer to ADP 3-0 for more information.) The variables are defined as follows:

- **Mission.** The mission is the task, together with the purpose, that clearly indicates the action to be taken and the reason. (Refer to JP 1-02 for more information.) Commanders analyze a mission in terms of specified tasks, implied tasks, and the commander's intent two echelons up.

- **Enemy.** This analysis includes not only the known enemy but also other threats to mission success. These include threats posed by multiple adversaries with a wide array of political, economic, religious, and personal motivations.

- **Terrain and weather.** Terrain and weather are natural conditions that profoundly influence operations. Terrain and weather are neutral; they favor neither side unless one is more familiar with— or better prepared to operate in—the physical environment. For tactical operations, terrain is analyzed using the five military aspects of terrain: observation, avenues of approach, key and decisive terrain, obstacles, and cover and concealment (OAKOC).

- **Troops and support available.** Troops and support available are the number, type, capabilities, and condition of available friendly troops and support. These include resources from joint, interagency, multinational, host nation, commercial (via contracting), and private organizations. They also include support provided by civilians.

- **Time available.** Time is critical to all operations. Controlling and exploiting it is central to initiative, tempo, and momentum. By exploiting time, commanders can exert constant pressure, control the relative speed of decisions and actions, and exhaust enemy forces.

- **Civil considerations.** Civil considerations reflect how the man-made infrastructure, civilian institutions, and attitudes and activities of civilian leaders, populations, and organizations within an area of operations influence the conduct of military operations. (Refer to ADP 3-0 for more information.) Civil considerations are areas, structures, capabilities, organizations, people, and events (ASCOPE).

THREATS

1-27. Threats facing U.S. forces today vary. They are not always enemy forces dressed in uniforms that are easily identified as foe, aligned on a battlefield and opposite U.S. forces. Threats are nation-states, organizations, people, groups, or conditions that can damage or destroy life, vital resources, or institutions.

1-28. Threats are described in four major categories or challenges: traditional, irregular, catastrophic, and disruptive. While helpful in describing the threats the Army is likely to face, these categories do not define the nature of the adversary. Adversaries may use any and all of these challenges in combination to achieve the desired effect against the U.S. (Refer to ADP 3-0 for more information.) The four threats are defined as follows:

- **Traditional.** States employing recognized military capabilities and forces in understood forms of military competition and conflict.
- **Irregular.** Opponent employing unconventional, asymmetric methods and means to counter traditional U.S. advantages.
- **Catastrophic.** Enemy that involves the acquisition, possession, and use of weapons of mass destruction and effects.
- **Disruptive.** Enemy using new technologies that reduce U.S. advantages in key operational domains.

Hybrid Threats

1-29. The term "hybrid threat" has recently been used to capture the seemingly increased complexity of operations and the multiplicity of actors involved. While the existence of innovative enemies is not new, today's hybrid threats demand that the company prepares for a range of possible threats simultaneously.

1-30. Hybrid threats are characterized by the combination of regular forces governed by international law, military tradition, and custom with irregular forces that are unregulated and as a result act with no restrictions on violence or targets for violence. This could include militias, terrorists, guerillas, and criminals. Such forces combine their abilities to use and transition between regular and irregular tactics and weapons. These tactics and weapons enable hybrid threats to capitalize on perceived vulnerabilities making them particularly effective.

DECISIVE ACTION

1-31. An Infantry rifle company operates in a framework of decisive action. ADP 3-0 provides a discussion of decisive action which includes the elements of offensive, defensive, and stability or civil support.

1-32. Army forces conduct decisive and sustainable land operations through the simultaneous combination of offensive, defensive, and stability operations (or defense support of civil authorities) appropriate to the mission and environment. Army forces conduct regular and irregular warfare against conventional and hybrid threats.

OFFENSE

1-33. Offensive operations are conducted to defeat and destroy enemy forces and seize terrain, resources, and population centers. They include movement to contact, attack, exploitation, and pursuit.

DEFENSE

1-34. Defensive operations are conducted to defeat an enemy attack, gain time, economize forces, and develop conditions favorable for offensive and stability tasks. These operations include mobile, area, and retrograde defense.

STABILITY OPERATIONS

1-35. Stability operations are military missions, tasks, and activities conducted outside the United States to maintain or reestablish a safe and secure environment, and to provide essential governmental services, emergency infrastructure reconstruction, and humanitarian relief. They include five tasks:

- Establish civil security.
- Establish civil control.
- Restore essential services.
- Support to governance.
- Support to economic and infrastructure development.

1-36. Homeland defense support of civil authorities represents Department of Defense support to U.S. civil authorities for domestic emergencies, law enforcement support, and other domestic activities, or from qualifying entities for special events. Tasks include providing support for—

- Domestic disasters.
- Domestic chemical, biological, radiological, nuclear, and high-yield explosives (CBRNE) incidents.
- Domestic civilian law enforcement agencies.
- Other designated support.

1-37. The simultaneous conduct of decisive action requires careful assessment, prior planning, and unit preparation as commanders shift their combinations of decisive action.

Note. For further information on decisive action refer to ADP 3-0.

MISSION-ESSENTIAL TASK LIST

1-38. A METL is a list of collective tasks a unit must perform successfully to accomplish its mission. (Refer to FM 7-0 for more information.)

1-39. To meet the demands of decisive action, the Headquarters, Department of the Army (HQDA) has standardized METL for brigades and above. This standardization ensures that like units deliver the same capabilities, and gives the Army the strategic flexibility to provide trained and ready forces to operational-level commanders. (See Figure 1-1.)

MISSION-ESSENTIAL TASK LIST DEVELOPMENT

1-40. The commander starts with reviewing the squadron METL and training guidance. The commander determines what collective tasks, battle drills, and leader tasks that support the squadron METL. The commander should include subordinate leaders in this task selection process because they must determine what individual tasks support the METL tasks. Based on the commander's analysis and identification of collective tasks that support the squadron METL, the commander determines a training focus that supports the squadron commanders training guidance. At the completion of METL development, the commander determines—

- Collective tasks that support the company METL.
- Individual tasks that support the METL tasks.
- Resources required for training to standards.

```
┌─────────────────────────────────────────────────────────────────┐
│  ┌──────────────────────────────────────────────┐                │
│  │  BN METL                                      │                │
│  │                                               │                │
│  │  17-6-1092, CONDUCT AN ATTACK                 │     ▲          │
│  └───────────┬──────────────────────────────────┘     │          │
│              ▼                                          │          │
│     ┌──────────────────────────────────────────────┐   │          │
│     │  CO METL                                      │   │          │
│     │                                               │   │          │
│     │  17-2-9001, CONDUCT AN ATTACK                 │   ▲          │
│     └───────────┬──────────────────────────────────┘   │          │
│                 ▼                                       │          │
│        ┌──────────────────────────────────────────┐    │          │
│        │  PLT-TEAM SUPPORTING COLLECTIVE TASK      │    │          │
│        │                                           │    │          │
│        │  07-3-9013, CONDUCT ACTION ON CONTACT     │    │          │
│        └───────────────────────────────────────────┘               │
└─────────────────────────────────────────────────────────────────┘
```

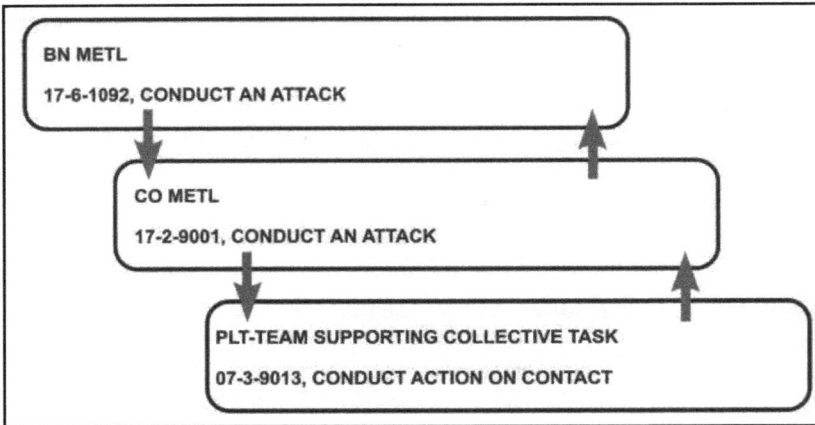

Figure 1-1. Collective tasks supporting higher unit METL

Commander's Analysis

1-41.   The commander initiates the METL development with an analysis of the BN METL and training guidance and he then identifies—

- The collective tasks, battle drills, and leader tasks the unit trains.
- The collective tasks the unit does not train and the risk for not training.
- An estimate of the time required to train.
- The conditions to train.
- Resources required.

Identify Collective Tasks

1-42.   The commander identifies the collective tasks, battle drills, and leader tasks to train and the estimated time required to train to proficiency. Additionally, the commander identifies those tasks the unit can accept risk for not training.

Identify the Conditions

1-43.   The commander gains an understanding of the OE that the company operates in and tries to replicate the training conditions if possible. The conditions determine what resources are needed to re-create the OE. The results of the troop commander's analysis are used to frame desired conditions in general terms.

1-44.   The commander, with input from the first sergeant (1SG), determines the scarce and unique resources needed to train the selected collective tasks and individual tasks in the conditions previously identified. The commander identifies those resources that require assistance from the BN commander to obtain. Identifying these requirements now gives the squadron commander and staff time for arranging and de-conflicting resources or finding alternatives.

Commander's Dialog

1-45. The higher commander approves the unit's METL. The approval normally occurs during the commander's dialog. The commander's dialog is a professional discussion between the company and BN commanders that set the expectations for developing a training plan. The company commanders' dialog is the culminating point of METL development. In general this event—

- Is conducted face-to-face.
- Sets expectations for planning company training.
- Identifies any unit training readiness problems or risks.
- Sets expectations for the development of the company training plan.
- Identifies the training risks for those tasks not trained.

1-46. Upon completion of this dialog, the company commander has the necessary products to publish the company METL and develop a training plan.

Implementation Guidance

1-47. The company commander issues a document to the company's officers and NCOs that summarizes the company commander's dialog with the BN commander. This is done primarily face-to-face. It provides the company officers and NCOs the necessary commander's guidance and training focus to develop platoon and squad training plans to achieve company METL proficiency.

PLANNING TRAINING

1-48. Training is formally planned at company and above levels. Training plans take the collective tasks to train and the assessment of proficiency in those tasks, and translate them into training events based on the commander's visualized end state. There are two types of training plans: long-range and short-range.

1-49. Commanders continuously assess the status (manning, equipping, and training) of the unit during training, and modify the long range training plan to build unit cohesion and achieve required METL proficiency as they move through the ARFORGEN force pools. (Refer to FM 7-0 for more information.)

TRAINING PRODUCTS

1-50.   Company commanders determine a training strategy for their unit and prepare training plans that enable the unit to be ready within the ARFORGEN process. Commanders develop training plans that enable them to attain proficiency in the METs needed to conduct decisive action under conditions in the OE.

1-51.   There are several training products available that the commander can use to train his unit to METL proficiency based on the readiness requirements. Each training product has been designed and developed within TRADOC to meet specific training needs. Commanders should consider the use of LVCG when considering training products. The following training products can be used throughout the training process of planning, preparation, execution, and assessment of unit training:

- Collective and individual tasks.
- Unit task lists (UTLs).
- Combined arms training strategies (CATSs).
- Warfighter training support packages (WTSPs).

INDIVIDUAL AND COLLECTIVE TASKS

1-52.   Both individual and collective tasks are performed during unit training in order to assess the proficiency of individuals and groups on their ability to perform the tasks to standard.

Note. This TC focuses on collective tasks and how they are used to support unit training and addresses individual tasks minimally.

Individual Tasks

1-53.   An individual task is a clearly defined, observable, and measurable activity accomplished by an individual. It is the lowest behavioral level in a job or duty that is performed for its own sake. An individual task supports one or more collective tasks or drills and often supports another individual task. Individual tasks can consist of both leader and staff tasks. The tasks are defined as follows:

- **Leader tasks.** This is an individual task (skill level 2 or higher) a leader performs that is integral to the performance of a collective task.
- **Staff tasks.** This is a clearly defined and measurable activity or action performed by a staff (collective) or a staff member (individual) of an organization who supports a commander in the exercise of unit mission command.

Collective Tasks

1-54. A collective task is a clearly defined, observable, and measurable activity or action that requires organized team or unit performance, leading to the accomplishment of a mission or function. Collective task accomplishment requires the performance to standard of supporting individual or collective tasks. Collective tasks can consist of shared and unique tasks. The tasks are defined as follows:

- **Shared.** A shared collective task is a collective task that applies to or is performed by more than one type of unit. Since the task, conditions, standards, task steps, and performance measures of shared collective tasks do not change, the collective task is trained and performed in the same way by all units that "share" the task. For example, Task 19-3-2406, *Conduct Roadblock and Checkpoint*, can be conducted by weapon and antiarmor companies.

- **Unique.** A unique collective task is a clearly defined, unit-specific collective task. For a collective task to be classified unique, no other unit or proponent (such as Infantry or engineers) may have the capability or requirement to perform the task. The designated proponent is solely responsible for the development and maintenance of a unique collective task.

1-55. Collective tasks are primarily performed in the operational domain, so the emphasis is on unit performance. Each collective task contains information that includes—

- **Assessment information.** Commanders can review the measures of performance and measures of success, and if the unit had performed those previously, and what the assessment was when performed. If an assessment was conducted, this assessment can provide needed information advising if the unit has performed the tasks and is considered trained (T), partially trained (P), or untrained (U).

- **General information.** These can include task title and warfighting function.

- **Task data, conditions, and standards.**

- **Task attributes.** These can include task trained at night, under mission-oriented protective posture (MOPP) conditions, and task steps.

- **Supporting information.** These can include products/references, individual tasks, drills, and collective tasks, and the prerequisite collective tasks.

Unit Task Lists

1-56. The unit task list (UTL) is a product of mission analysis that identifies all of the collective tasks (shared and unique) that a unit is organized, manned and equipped to conduct. The UTL is produced for each unit with a table of organization and equipment (TOE)/modified TOE (MTOE) or table of distribution and allowance (TDA).

1-57. The UTL can include existing collective tasks, or collective tasks identified to be designed and developed. The UTL also provides the baseline for a unit combined arms training strategy (CATS). A training developer creates the UTL by linking collective tasks to those missions identified on the TOE. This process ensures that units train the appropriate tasks to required proficiency levels.

UTL Locations

1-58. An assembled UTL is located in Appendix A. The UTL is also maintained and accessed within the Digital Training Management System (DTMS).

Digital Training Management System

1-59. The DTMS is a web-based training management system that allows the unit to conduct mission and METL development, training planning and management, and track unit training by implementing the doctrine, tactics, techniques, and procedures outlined in FM 7-0.

Combined Arms Training Strategy

1-60. The combined arms training strategy is the Army's overarching strategy for the current and future training of the force. It describes how the Army trains the total force to standard in the institution, unit, and through self-development. It also identifies, quantifies, and justifies the training resources required to execute the training. Unit CATSs are built using the unit missions and the UTL, and are designed to reflect the METL. CATSs have replaced mission training plans (MTPs). CATSs provide a training path with recommendations of what and who to train. CATSs support the unit METL training and are synchronized with ARFORGEN.

Types of CATS

1-61. Combined arms training strategies are based on the unit's TOE mission (that support the METL), employment, capabilities, and functions. There are two types of CATSs: unit and functional.

Unit CATS

1-62. Unit CATSs are TOE-based and unique to a unit type. Unit CATS development considers organizational structure, METL, and doctrine to organize the unit's collective tasks in a strategy that provides a path for achieving task proficiency.

1-63. A CATS consists of a menu of task selections that provide a base strategy for unit commanders to plan, prepare, and assess training to provide a flexible training strategy. CATSs are also designed to train a capability with supporting training events and resources. The events are designed to be trained in a logical sequence, starting with the lowest echelon and adding echelons as the events get progressively more complex. The culminating, or run event, for a CATS is usually the highest level event designed to train and/or evaluate the entire unit.

1-64. Unit CATSs provide commanders training strategies from which they develop their unit training plan to achieve collective task proficiency, as well as support the ARFORGEN phases. These strategies are flexible and are not intended to constrain commanders but rather provide them with a menu of core mission/capabilities-based training events. They provide commanders with a method to train all tasks that a unit is designed to perform and estimate the required resources to support event-driven training. Unit CATS provide commanders with tools to plan, prepare for, and evaluate unit training.

Functional CATS
1-65. Functional CATSs address a functional capability common to multiple units and echelons and they supplement unit CATS. Strategies may be based on missions or functions performed by most units that are not unique to a specific unit type, or they may be developed to train warfighting functions or operational themes that were not incorporated into unit CATS. Two examples of functional CATS are mission command (currently listed in CATS as command and control), and force protection. Functional CATSs contain most of the same data elements as unit CATS.

Task Selections
1-66. Task selections describe specific capabilities/missions, and include collective tasks that support developing those capabilities. Frequency of training and types of events that can be used to train the capability are also recommended.

1-67. Task selections are usually trained utilizing a series of crawl-walk-run events. Crawl-walk-run events provide options to commanders to accommodate training at the appropriate level of difficulty based on their training readiness assessment. Each event provides recommendations for who and how to train, and resources required to support that training.

1-68. The commander can consolidate the necessary collective tasks to be trained to support the METs, which helps to determine the time and resources needed to train these tasks to proficiency. A matrix showing the CATS task selections used within the weapons and antiarmor company that support the task groups of the unit's METL are located in Appendix C.

1-69.   Commanders review applicable task selections in CATSs to develop select events that support externally directed events. Task selections also identify training-gate events for key training events. Additionally, CATSs provide the recommended frequency and sequence for scheduling training events.

Training Events

1-70.   Commanders organize collective and individual tasks into standard Army training events. When conducted, they adhere to the principles of training mentioned earlier in this chapter.

1-71.   The commander can also develop training events internally, such as classes, sergeants time training (STT), field training exercises (FTXs), situation training exercises (STXs), and combined arms live fire exercises (CALFEXs), when using the crawl-walk-run training path provided within CATS. Commanders can utilize the training gates developed for assessment of unit proficiency of each training event.

1-72.   The commander may create different versions of unit training plans using the CATS. A unit's progress through its training plan is based on time available and the commander's assessment of task proficiency using the doctrinal process of assessing training, missions, and METs while preparing or updating unit training plans.

CATS Locations

1-73.   Combined arms training strategies are available digitally from both DTMS and the ATN. In digital format, the CATS provides numerous links to training materials, which can assist the commander and unit training managers to develop the commander's plan and to conduct training.

Warfighter Training Support Packages

1-74.   The warfighter training support package (WTSP) is a complete, stand alone, exportable training package integrating training products and materials needed to train one or more collective tasks and supporting critical individual tasks. WTSPs are task-based information packages that provide structured situational training scenarios for LVCG unit or institutional training.

1-75.   Warfighter training support packages assist commanders in training their unit's METL. This is accomplished by basing the WTSP on a revised Caspian Sea Scenario for differing echelons. Each WTSP includes materials to support planning, preparing, executing and assessing training exercises at respective echelons. The WTSP can aid the commander throughout the training management of their unit during various training exercises.

Warfighter Training Support Packages Locations

1-76. Warfighter training support packages are exportable for use by the unit. Unlike CATS and UTLs that can be accessed through DTMS or the ATN. The company team WTSPs is located within the Maneuver Center of Excellence Collective Training Branch website on Army Knowledge Online (AKO). To access this website:

- Log into AKO.
- Copy and paste the Web address, (https://www.us.army mil/suite/grouppage/130823), into the address bar.
- Select enter.
- Select desired WTSP.

LIVE, VIRTUAL, CONSTRUCTIVE, AND GAMING TRAINING

1-77. Company commanders can use LVCG training to enhance training, replicate battlefield conditions, balance resources, and sustain readiness. Commanders consider each of these to dictate the degree of simulation they plan for their unit during training events. Utilizing simulations within their unit training enables commanders to maximize many of the principles of training and to manage scarce resources.

LIVE

1-78. Live training is executed in field conditions using tactical equipment. It involves real people operating real systems. Live training may be enhanced by training aids, devices, simulators, and simulations (TADSS) and tactical engagement simulation (TES) to simulate combat conditions. Use of simulation is used to improve a unit's marksmanship caliber.

1-79. The Initial Homestation Instrumented Training System (I-HITS) provides position location and weapons effects data for real-time exercise monitoring and AAR capability. The Instrumentable-Multiple Integrated Laser Engagement System (I-MILES) has replaced the basic Multiple Integrated Laser Engagement System (MILES) that is currently fielded. This new system provides the real-time casualty effects necessary for tactical engagements training in direct-fire, force-on-force, and instrumented training scenarios.

Note. No enhanced training can duplicate firing live rounds.

VIRTUAL

1-80. Virtual training is executed using computer-generated battlefields in simulators with the approximate characteristics of tactical weapon systems and vehicles. It exercises motor control, decision making, and communication skills. Sometimes called "human-in-the-loop training," it involves real people operating simulated systems. Soldiers being trained practice the skills needed to operate actual equipment.

1-81. An example of virtual training is the close combat tactical trainer (CCTT). This system is equipped with the latest Force XXI Battle Command Brigade and Below (FBCB2) in support of training the digital force. Dismounted Soldier is part of the CCTT program. It provides the capability to train Soldiers and units in all three elements of decisive action described in ADP 3-0.

CONSTRUCTIVE

1-82. Constructive training uses computer models and simulations to exercise command and staff functions. It involves simulated people operating simulated systems.

1-83. Constructive training can be conducted by units from platoon through echelons above corps. A command post (CP) exercise is an example of constructive training. Other examples are Multi-User Online Virtual Exercise (MOVE) and hands-on-trainer (HOT).

GAMING

1-84. Gaming is the use of technology employing commercial or government off-the-shelf, multigenre games in a realistic, semi-immersive environment to support education and training. The military uses gaming technologies to create capabilities to help train individuals and organizations.

1-85. Gaming can enable individual, collective, and multiechelon training. It can operate in a stand-alone environment or be integrated with live, virtual, or constructive enablers. Employed in a realistic, semi-immersive environment, gaming can simulate operations and capabilities. An example of fully interactive, three-dimensional gaming is Virtual Battlespace System 2 (VBS2), a mission rehearsal tool for Soldiers to practice tactics, techniques, and procedures in a synthetic environment prior to conducting an actual mission. Another example is the DARWARS Ambush designed for convoy operation training, platoon-level mounted Infantry tactics, and dismounted operations. Urban Simulation (URBANSIM) and Command Post of the Future (CPOF) are also virtual training gaming aids.

Chapter 2

Crosswalks and Outlines

This chapter provides the weapons and antiarmor company commander a METL crosswalk, a CATS task selection to METL matrix, and the METL supporting collective task training and evaluation outlines (T&EOs). Each item assists the commander and leaders with training within the company.

SECTION I – METL CROSSWALK

2-1.   Table 2-1 contains the primary references for conducting decisive actions - offensive, defensive, and stability operations (or defense support of civil authorities). For more information on how to plan, prepare, and execute the collective tasks and drills in this chapter refer to the references in Table 2-1.

2-2.   Commanders focus their training efforts on training collective tasks that support the BN METL. One of the many responsibilities of the commander is to determine which tasks to train. This crosswalk is a tool the commander can use as a starting point for selecting the supporting collective task-to-BN METL. The supporting collective task to the BN METL crosswalk is an example developed by the Directorate of Training and Doctrine, MCoE. (See Table 2-2.) The crosswalk identifies tasks that support the BN mission-essential task (MET). Supporting collective tasks that support the BN METL are aligned on the left side of the matrix. The "X" identifies the supporting collective tasks that support the MET.

Table 2-1. Primary references for decisive actions

Decisive Action	References
Offensive, Defensive, and Security Operations	FM 3-21.12, *The Infantry Weapons Company* FM 3-21.91, *Tactical Employment of Antiarmor Platoons and Companies* FM 3-90, *Tactics*

Table 2-1. Primary references for decisive actions (continued)

Decisive Action	References
Stability Operations	FM 3-07, *Stability Operations*
Defense Support of Civil Authorities	FM 3-28, *Civil Support Operations*

Table 2-2. Example of weapons and antiarmor company METL to task crosswalk

Weapons and Antiarmor Company		METs and Task Groups			
		Attack	*Defend*	*Security*	*Stability*
Task Number	*Task Title*	*Movement to Contact*	*Deliberate Attack*	*Area Defense*	*Screen*	*Area Security*	*Public Order & Safety*
07-2-1090	Conduct a Movement to Contact (Platoon-Company)	X					
07-2-1256	Conduct an Attack by Fire (Platoon-Company)	X	X				
07-2-1252	Conduct an Antiarmor Ambush (Platoon-Company)		X				
07-2-1324	Conduct Area Security (Platoon-Company)					X	X
07-2-3000	Conduct Support by Fire (Platoon-Company)	X	X				
07-2-3036	Integrate Indirect Fire Support (Platoon-Company)					X	X

Table 2-2. Example of weapons and antiarmor company METL to task crosswalk (continued)

Weapons and Antiarmor Company		METs and Task Groups			
		Attack	*Defend*	*Security*	*Stability*
Task Number	*Task Title*	*Movement to Contact* / *Deliberate Attack*	*Area Defense* / *Screen*	*Area Security*	*Public Order & Safety*
07-2-4054	Secure Civilians During Operations (Platoon-Company)			X	X
07-2-5027	Conduct Consolidation and Reorganization (Platoon-Company)	X X X		X	
07-2-9001	Conduct an Attack (Platoon-Company)	X			
07-2-9002	Conduct a Bypass (Platoon-Company)	X X			
07-2-9003	Conduct a Defense (Platoon-Company)	X X		X	
07-2-9004	Conduct a Delay (Platoon-Company)		X		
07-2-9006	Conduct a Passage of Lines as the Passing Unit (Platoon-Company)	X X			
07-2-9007	Conduct a Passage of Lines as the Stationary Unit (Platoon-Company)		X		

Table 2-2. Example of weapons and antiarmor company METL to task crosswalk (continued)

Weapons and Antiarmor Company		METs and Task Groups					
		Attack		Defend	Security		Stability
Task Number	Task Title	Movement to Contact	Deliberate Attack	Area Defense	Screen	Area Security	Public Order & Safety
07-2-9009	Conduct a Withdrawal (Platoon-Company)			X			
07-2-9012	Conduct a Relief in Place (Platoon-Company)			X	X		
08-2-0003	Treat Casualties	X	X	X	X	X	X
08-2-0004	Evacuate Casualties	X	X	X	X	X	X
17-2-9225	Conduct a Screen (Platoon-Company)			X		X	
19-3-2406	Conduct Roadblock and Checkpoint	X	X	X	X	X	
63-2-4546	Conduct Logistics Package (LOGPAC) Support			X		X	X
03-2-9224	Conduct Operational Decontamination	X	X	X		X	
34-5-0471	Support Company Level Intelligence, Surveillance, and Reconnaissance (ISR)	X	X	X	X	X	X

Table 2-2. Example of weapons and antiarmor company METL to task crosswalk (continued)

Weapons and Antiarmor Company		METs and Task Groups					
		Attack		*Defend*		*Security*	*Stability*
Task Number	Task Title	*Movement to Contact*	*Deliberate Attack*	*Area Defense*	*Screen*	*Area Security*	*Public Order & Safety*
34-5-0470	Provide Situational Awareness of the Company Area of Operations	X	X	X	X	X	X
34-5-0472	Provide Intelligence Support Team Input to Targeting	X	X	X	X	X	X

SECTION II – TRAINING AND EVALUATION OUTLINES

INTRODUCTION

2-3. This section provides the supporting collective tasks in the form of T&EOs. All T&EOs support unit missions, and individual T&EOs may support multiple missions.

2-4. Leaders and Soldiers within the unit can use them as a reference on how to perform a task. Commanders and leaders may use them to identify subordinate unit supporting tasks. Observers or evaluators can use them to record and document the unit's task performance.

STRUCTURE

2-5. Each T&EO provides the task conditions and standards. They also provide a series of task steps and performance measures that serve as a logical guide for performing the task. The task steps are generally sequential, but they may be performed concurrently, or even omitted, based on the mission variables of METT-TC. The unit's ability to accomplish the task steps and performance measures indicates whether or not it is executing the task to standard. Table 2-1 lists METL tasks by METs and task groups, with task title and numbers to that specific T&EO.

FORMAT

2-6.   Each T&EO displayed in this TC consists of the following:

- **Task.** This is a description of the action to be performed by the unit, and provides the task number.
- **References.** These are in parenthesis following the task number. The reference that contains the most information (primary reference) about the task is listed first.
- **Condition.** The condition is a written statement of the situation or environment in which the unit is to do the collective task.
- **Task standard.** States the performance criteria that a unit must achieve to successfully execute the task. This overall standard should be the focus of training and understood by every Soldier. The trainer or evaluator determines the unit's training status using performance observation measurements (where applicable) and his judgment. The unit must be evaluated in the context of the METT-TC conditions. These conditions should be as similar as possible for all evaluated elements. This establishes a common base line for unit performance.
- **Task steps and performance measures.** This is a listing of actions that is required to complete the task. These actions are stated in terms of observable performance for evaluating training proficiency. The task steps are arranged sequentially along with supporting individual tasks and their reference. Leader tasks within each T&EO are indicated by an asterisk (*). Under each task step are listed the performance measures that must be accomplished to correctly perform the task step. If the unit fails to correctly perform one of these task steps to standard, it has failed to achieve the overall task standard.
- **GO/NO-GO column.** This column is provided for annotating the unit's performance of the task steps. When assessing training, evaluate each performance measure for a task step and place an "X" in the appropriate column. A major portion of the performance measures must be marked a "GO" for the task step to be successfully performed.

- **Supporting collective tasks.** This is a clearly defined, discrete, and measurable activity, action, or event (for example, task) that requires organized or unit performance, and leads to accomplishment of a mission.

USE

2-7. The T&EOs can be used to train or evaluate a single task. Several T&EOs may be used by an observer controller as an evaluation outline or by a commander as a training outline.

TASK: Conduct a Movement to Contact (Platoon-Company) (07-2-1090)

(FM 3-21.10)   (FM 3-21.8)

CONDITIONS: The unit conducts operations as part of a higher headquarters (HQ) and receives an operation order (OPORD) or fragmentary order (FRAGO) to conduct a movement to contact to gain or regain contact with the enemy. Communications have been established, and information is being passed according to the unit standing operating procedures (SOPs). The unit receives guidance on the rules of engagement (ROE). Coalition forces and noncombatants may be present in the area of operational environment. Some iterations of this task should be conducted during limited visibility conditions and should be performed in mission-oriented protective posture 4 (MOPP 4).

STANDARDS: The unit conducts the movement to contact according to the SOPs, the order, and/or the commander's guidance. The unit leader selects the proper technique for conducting the movement to contact based on the anticipated enemy situation. The unit finds, fixes, develops the situation, and finishes the enemy. The unit moves not later than the time specified in the order, reports required intelligence information, and complies with the ROE.

TASK STEPS AND PERFORMANCE MEASURES	GO	NO-GO

PLAN
*1. Unit leaders gain and/or maintain situational understanding using available communications equipment, maps, intelligence summaries, situation reports (SITREPs), and other available information sources. Intelligence sources include company intelligence support team (CoIST), human intelligence (HUMINT), signal intelligence (SIGINT), and imagery intelligence (IMINT) to include unmanned aircraft systems (UASs), and unattended ground sensors (UGSs).
*2. Unit leaders receive an OPORD or FRAGO and issue a warning order (WARNO) to include at a minimum—
    a.   The mission or nature of the movement to contact.
    b.   The time and place for issuing the OPORD.
    c.   Units or elements participating in the movement to contact.
    d.   Specific tasks not addressed by unit SOPs.
    e. The timeline for the movement to contact.		

TASK STEPS AND PERFORMANCE MEASURES	GO	NO-GO

*3. Unit leaders confirm friendly and enemy situations. They take the following actions:
    a. Receive an updated report showing the location of forward and adjacent friendly elements, if applicable.
    b. Receive an updated enemy situational template for added fratricide prevention and increased force protection, if applicable.
    c. Clarify priority intelligence requirement (PIRs).
    d. Confirm any changes to the HQ and unit task or purpose.
    e. Confirm any changes to the scheme of maneuver.
4. Unit leaders perform the following fundamentals:
    a. Focus all efforts on finding the enemy.
    b. Make initial contact with the smallest force possible, consistent with protecting the force.
    c. Make initial contact with small, mobile, self-contained forces to avoid decisive engagement of the main body on ground chosen by the enemy. (This allows the commander maximum flexibility to develop the situation.)
    d. Task-organize the force and use movement formations to deploy and attack rapidly in any direction.
    e. Keep forces within supporting distances to facilitate a flexible response.
    f. Maintain contact regardless of the course of action (COA) adopted once contact is gained.
*5. Unit leaders conduct troop-leading procedures. They take the following actions:
    a. Conduct a map reconnaissance. Take the following actions:
        (1) Identify tentative rally points, if required.
        (2) Identify likely enemy avenues of approach.
    b. Coordinate indirect fire support and or close air support, if available.
    c. Conduct direct fire planning.
    d. Plan the integration of direct and indirect fires according to HQ's fire support plan.
    e. Select the proper technique below for conducting the movement to contact, if not directed by HQ:
        (1) Search-and-attack technique.
        (2) Cordon and search.		

TASK STEPS AND PERFORMANCE MEASURES	GO	NO-GO
f. Organize the unit as needed to accomplish the mission and/or compensate for combat losses.		
g. Plan continuous operations if required.		
h. Plan and coordinate support.		
i. Determine the requirement for patrol bases/assembly area.		
j. Determine linkup requirements.		
k. Determine if movement technique is based on factors of mission, enemy, terrain and weather, troops and support available, time available, and civil considerations (METT-TC).		
l. Determine how key weapons will be employed.		
m. Confirm fire control measures and engagement criteria.		
n. Address actions on contact with the enemy.		
o. Consider enemy capabilities, likely COAs, and specific weapons capabilities to understand the threat and ensure the security of the unit.		
p. Coordinate with adjacent units as required.		
q. Coordinate passage of lines if required.		
r. Decide what formations the unit uses to enter and move in the zone or area of operations, and what the contingency plans are.		
s. Conduct reconnaissance as required. (There may not be enough time to reconnoiter extensively to locate the enemy.) Take the following actions:		
(1) Confirm the most likely enemy location.		
(2) Adjust the plan based on updated intelligence and reconnaissance effort.		
(3) Update the enemy situation.		
(4) Disseminate updated reports (if applicable), overlays, and other pertinent information.		
PREPARE		
*6. Unit leaders issue clear and concise orders.		
7. The unit conducts a rehearsal.		
*8. Unit leaders issue FRAGOs, as needed, to address changes to the plan identified during the rehearsal.		
*9. Unit leaders coordinate and/or synchronize actions of subordinate elements.		

TASK STEPS AND PERFORMANCE MEASURES	GO	NO-GO
*10. Unit leaders use FRAGOs as needed to redirect actions of subordinate elements. EXECUTE 11. The unit executes search-and-attack for one or more of the following purposes: a. To protect the force—prevent the enemy from massing to disrupt or destroy friendly military or civilian operations, equipment, property, and key facilities. b. To collect information—gain information about the enemy and the terrain to confirm the enemy course of action (ECOA) predicted by the intelligence preparation of the battlefield (IPB) process. Help generate situational awareness (SA) for the company and HQ. c. To destroy the enemy and render enemy units in the AO combat ineffective. d. To deny the area—prevent the enemy from operating unhindered in a given area, such as in a base camp or for logistics support. 12. The unit executes cordon and search based on the situation. (See Collective Task, Conduct a Cordon and Search (Platoon-Company) 07-2-9051.) *13. Unit leaders synchronize element actions. ASSESS 14. The unit consolidates and reorganizes as needed. 15. The unit continues operations as directed. *indicates a leader task step.		

SUPPORTING INDIVIDUAL TASKS

Task Number	Task Title
061-283-6003	Adjust Indirect Fire
071-410-0010	Conduct a Leaders Reconnaissance
071-420-0005	Conduct the Maneuver of a Platoon
171-620-0094	Conduct Consolidation and Reorganization Activities at Company-Troop Level
071-326-5503	Issue a Warning Order
071-326-5505	Issue an Operation Order at the Company, Platoon, or Squad Level
071-326-5502	Issue a Fragmentary Order
171-121-4045	Conduct Troop Leading Procedures

SUPPORTING COLLECTIVE TASKS

Task Number	Task Title
07-2-1342	Conduct Tactical Movement (Platoon-Company)
07-2-1450	Secure Routes (Platoon-Company)
07-2-3027	Integrate Direct Fires (Platoon-Company)
07-2-3036	Integrate Indirect Fire Support (Platoon-Company)
07-2-5009	Conduct a Rehearsal (Platoon-Company)
07-2-5027	Conduct Consolidation and Reorganization (Platoon-Company)
07-2-5063	Conduct Composite Risk Management (Platoon-Company)
07-2-6063	Maintain Operations Security (Platoon-Company)
07-2-9002	Conduct a Bypass (Platoon-Company)
07-2-9006	Conduct a Passage of Lines as the Passing Unit (Platoon-Company)
07-2-9014	Occupy an Assembly Area (Platoon-Company)
07-3-1072	Conduct a Disengagement
07-3-9013	Conduct Action on Contact
07-3-9017	Conduct Actions at Danger Areas
08-2-0003	Treat Casualties
08-2-0004	Evacuate Casualties
07-2-9051	Conduct a Cordon and Search (Platoon-Company)
07-2-3000	Conduct a Support by Fire (Platoon-Company)
07-2-1256	Conduct a Support by Fire (Platoon-Company)

SUPPORTING BATTLE/CREW DRILLS

Task Number	Task Title
07-3-D9501	React to Contact (Visual IED, Direct Fire [includes RPG])

TASK: Conduct an Attack by Fire (Platoon-Company) (07-2-1256)

(FM 3-21.10)　　(FM 3-21.8)

CONDITIONS: The unit conducts operations as part of a higher headquarters (HQ) and receives an operation order (OPORD) or fragmentary order (FRAGO) to conduct an attack by fire. The unit is assigned a battle position (BP) and a sector of fire, an engagement area (EA), or an axis of advance and objective. The enemy may be stationary or moving. Communications are established, and information is passed according to unit standing operating procedures (SOPs). The unit receives guidance on the rules of engagement (ROE). Coalition forces and noncombatants may be present in the operational environment. Some iterations of this task should be conducted during limited visibility conditions and performed in mission-oriented protective posture 4 (MOPP 4).

TASK STANDARDS: The unit conducts the attack by fire according to the SOPs, the order, and/or the commander's guidance.

TASK STEPS AND PERFORMANCE MEASURES	GO	NO-GO
PLAN		
*1. Unit leaders gain and/or maintain situational understanding using available communications equipment, maps, intelligence summaries, situation reports (SITREPs), and other available information sources. Intelligence sources include company intelligence support team (CoIST), human intelligence (HUMINT), signal intelligence (SIGINT), and imagery intelligence (IMINT) to include unmanned aircraft systems (UASs) and unattended ground sensors (UGSs).		
*2. Unit leaders receive an OPORD or FRAGO and issue a warning order (WARNO) to include at a minimum—		
a.　The mission or nature of the attack by fire.		
b.　The time and place for issuing the OPORD.		
c.　Units or elements participating in the attack by fire.		
d.　Specific tasks not addressed by unit SOPs.		
e.　The timeline for the attack by fire.		
*3. Unit leaders plan using troop-leading procedures (TLPs). They take the following actions:		
a.　Conduct a map reconnaissance. Take the following actions:		

TASK STEPS AND PERFORMANCE MEASURES	GO	NO-GO
(1) Identify attack by fire (ABF) and sector of fire or EA.		
(2) Identify likely enemy avenues of approach or axis of advance.		
(3) Identify routes to and from the ABF positions.		
(4) Identify tentative target reference points (TRPs).		
(5) Mark tentative dismount points on maps if mounted.		
b. Plan and coordinate indirect fire support and or close air support if available.		
c. Organize the unit as necessary to accomplish the mission and or compensate for combat losses.		
d. Address actions on chance contact with the enemy.		
e. Disseminate applicable reports, overlays, and other pertinent information.		
f. Plan control measures for lifting or shifting direct and indirect fires.		
g. Ensure observers are positioned to adjust indirect fires if applicable.		
PREPARE		
*4. Unit leaders or designated representatives conduct a reconnaissance. They take the following actions:		
a. Select ABF positions that allow the unit to effectively engage the enemy and that provide adequate cover and concealment.		
b. Establish and leave security at the ABF position.		
c. Designate engagement criteria, rate of fire, weapons distribution and engagement priorities by weapons system.		
d. Update intelligence information.		
e. Return to the unit position.		
*5. Unit leaders adjust the ABF plan, if necessary, based on updated intelligence.		
*6. Unit leaders issue orders and instructions to include ROE.		
7. The unit conducts a rehearsal.		
*8. Unit leaders issue FRAGOs, as necessary, to address changes to the plan identified during the rehearsal.		

TASK STEPS AND PERFORMANCE MEASURES	GO	NO-GO

9. The unit conducts tactical movement to ABF position. It takes the following actions:
     a.   Employs appropriate formation and movement technique.
     b.   Uses covered and concealed routes to prevent the enemy from effectively engaging the unit.
     c.   Orients weapon systems to provide 360-degree security during movement.
10. The unit occupies ABF position. It takes the following actions:
     a.   Confirms the position meets the following tactical considerations:
          (1) Allows the unit to place effective fires on the enemy.
          (2) Facilitates weapon standoff.
          (3) Is located on terrain affording cover and concealment.
     b.   Conducts hasty occupation of the ABF position.
     c.   Designates TRPs, sectors of fire, and tentative firing positions.
     d.   Begins scanning sectors of fire as designated by unit leaders.
EXECUTE
11. The unit observes the designated engagement areas or sectors of responsibility. It takes the following actions:
     a.   Detects all enemy entering the area.
     b.   Notifies supported, flanking, and higher units of detected enemy as required.
12. The unit executes the attack by fire against the enemy. It takes the following actions:
     a.   Acquires, suppresses, and/or destroys all identified enemy elements using appropriate weapon systems.
     b.   Calls for and adjusts indirect fires to block and or destroy the enemy.
     c.   Maneuvers to alternate positions as necessary to maintain effective fires on the enemy or to maintain survivability.
     d. Shifts, refocuses, and redistributes direct fires as necessary to destroy the enemy.		

TASK STEPS AND PERFORMANCE MEASURES	GO	NO-GO

*13.Unit leaders direct the attack by fire until all enemy elements are destroyed, fixed, or suppressed or the order to lift or shift fires is received. They take the following actions:
    a.   Focus and distribute direct fires and shifts; refocus and redistribute fires to maintain suppression of the enemy or to destroy enemy forces.
    b.   Shift indirect fires to suppress or destroy enemy vehicles or positions.
    c.   Lift fires to facilitate the movement of friendly elements or when target effects are achieved.
    d.   Issue additional FRAGOs to direct or task subordinate elements as required.
    e.   Order a cease-fire once the enemy is destroyed or on order from the commander.
    f.   Send spot reports (SPOTREPs), update SITREPs, and make recommendations to the higher commander as required.
ASSESS
14. The unit consolidates and reorganizes as needed.
15. The unit continues operations as directed.
*indicates a leader task step		

SUPPORTING INDIVIDUAL TASKS

Task Number	Task Title
171-610-0001	Perform a Map Reconnaissance
061-283-6003	Adjust Indirect Fire
071-410-0010	Conduct a Leaders Reconnaissance
071-326-5503	Issue a Warning Order
071-326-5502	Issue a Fragmentary Order
071-326-5505	Issue an Operation Order at the Company, Platoon, or Squad Level
171-121-4054	Conduct Troop-Leading Procedures
071-326-5630	Conduct Movement Techniques by a Platoon
071-420-0005	Conduct the Maneuver of a Platoon

SUPPORTING COLLECTIVE TASKS

Task Number	Task Title
07-2-1342	Conduct Tactical Movement (Platoon-Company)
07-2-1396	Employ Obstacles (Platoon-Company)
07-2-3000	Conduct Support by Fire (Platoon-Company)

07-2-3027	Integrate Direct Fires (Platoon-Company)
07-2-3036	Integrate Indirect Fire Support (Platoon-Company)
07-2-5009	Conduct a Rehearsal (Platoon-Company)
07-2-5027	Conduct Consolidation and Reorganization (Platoon-Company)
07-2-5063	Conduct Composite Risk Management (Platoon-Company)
07-2-6063	Maintain Operations Security (Platoon-Company)
07-2-9006	Conduct a Passage of Lines as the Passing Unit (Platoon-Company)
07-3-9013	Conduct Action on Contact
07-3-9016	Establish an Observation Post
08-2-0003	Treat Casualties
08-2-0004	Evacuate Casualties

SUPPORTING BATTLE/CREW DRILLS

Drill Number	Drill Title
07-3-D9501	React to Contact (Visual, IED, Direct Fire [includes RPG])
05-3-D0016	Conduct the 5 Cs

TASK: Conduct an Antiarmor Ambush (Platoon-Company) (07-2-1252)

(FM 3-21.91) (FM 3-21.10)

CONDITIONS: The unit conducts operations as part of a larger force and receives an operation order (OPORD) or fragmentary order (FRAGO) directing the conduct of an antiarmor ambush. The enemy conducts mounted operations, along lines of communications or on natural lines of drift. All needed personnel and equipment are available. The unit has communications with higher, adjacent, and subordinate elements. The unit has been provided guidance on the rules of engagement (ROE). Coalition forces and noncombatants may be present in the operational environment. Some iterations of this task should be performed in mission-oriented protective posture 4 (MOPP 4).

STANDARDS: The unit emplaces the ambush according to standing operating procedures (SOPs), the order, and/or commander's guidance. The unit surprises and engages the enemy main body and destroys the vehicles in the kill zone. The unit destroys enemy equipment in the kill zone based on the commander's intent.

TASK STEPS AND PERFORMANCE MEASURES	GO	NO-GO

PLAN
*1. Unit leaders gain and or maintain situational understanding using available communications equipment, maps, intelligence summaries; situation reports (SITREPs) and other available information sources. Intelligence sources include company intelligence support team (CoIST), human intelligence (HUMINT), signal intelligence (SIGINT), and imagery intelligence (IMINT) to include unmanned aircraft systems (UASs) and unattended ground sensors (UGSs).
*2. The unit leader receives an OPORD or FRAGO and issues warning order (WARNO) to include at a minimum—
   a. The mission or nature of the ambush.
   b. The time and place for issuing the OPORD.
   c. Units or elements participating in the ambush.
   d. Specific tasks not addressed by unit SOPs.
   e. The timeline for the ambush.
*3. The unit leader plans using troop-leading procedures (TLPs). He takes the following actions:
   a. Conducts a map reconnaissance.		

TASK STEPS AND PERFORMANCE MEASURES	GO	NO-GO
b. Determines tentative positions for the security element, support element, and assault element.		
c. Develops criteria for initiation of the ambush.		
d. Determines signals for initiating the ambush, cease fire, and withdrawal from the ambush site.		
e. Determines actions in the kill zone.		
PREPARE		
4. The unit moves to hide positions vicinity of the ambush site. It takes the following actions:		
a. Conducts passage of lines through friendly units (if required) and then moves to designated hide positions.		
b. Occupies the hide positions according to the OPORD, FRAGO or unit SOPs.		
c. Plans withdrawal routes.		
*5. The unit leader conducts a leader's reconnaissance to confirm or modify the plan. He takes the following actions:		
a. Ensures the reconnaissance party moves undetected.		
b. Selects the security element positions. Takes the following actions:		
(1) Provides early warning of enemy approach into the engagement area or toward any element.		
(2) Designates the engagement techniques, if not according to OPORD, FRAGO, or SOPs.		
(3) Provides covered and concealed positions.		
(4) Allows the security element to secure the unit, prevent the enemy escaping from the kill zone, and prevent reinforcement of the enemy in the kill zone.		
c. Selects the kill zone. Takes the following actions:		
(1) On a canalized armor/mechanized avenue of approach.		
(2) Restricts enemy maneuver within the kill zone.		
d. Selects the support position(s). Takes the following actions:		
(1) Positions mortars (if available) to provide fire into the kill zone and to support the security element with indirect fire.		
(2) Permits observation and effective fires into the kill zone.		
(3) Provides cover and concealment.		
(4) Ensures positions are protected by obstacles (natural or man-made).		

TASK STEPS AND PERFORMANCE MEASURES	GO	NO-GO
e. Identifies target reference points (TRPs), element engagement areas (EAs), and direct fire responsibilities.		
f. Identifies trigger point(s) and decision points.		
g. Leaves a surveillance team at the ambush site.		
h. Returns to hide positions undetected.		
*6. The unit leader confirms or modifies the plan based on the reconnaissance. He takes the following actions:		
a. Reviews movement to, occupation of and actions at the ambush site.		
b. Briefs ROE and provide guidance on acceptable collateral damage limits.		
c. Allows time for subordinate leaders to brief elements of any modifications to the plan.		
d. Ensures understanding on movement to and occupation of the ambush site.		
e. Reports readiness to execute to higher headquarters (HQ).		
*7. The security element leader moves his element from the hide positions and occupies security positions. He takes the following actions:		
a. Moves tactically to the positions.		
b. Designates individual sectors of observation and fire.		
c. Directs the emplacement of obstacles and early warning devices, as the enemy situation dictates.		
d. Checks, if possible, the positions from the enemy side to verify concealment.		
e. Positions weapon systems for best control.		
f. Verifies status of all sensors, target acquisition, and night vision devices.		
g. Maintains communication with the unit leader.		
h. Ensures the occupation and establishment of the positions is undetected by the enemy.		
i. Informs the unit leader when the positions are established.		
8. Assault and any support elements occupy their BPs. They take the following actions:		
a. Move tactically to the positions.		
b. Designate the individual sectors of observation and fire, and prepare sector sketches.		

TASK STEPS AND PERFORMANCE MEASURES	GO	NO-GO
c. Emplace obstacles, and early warning devices, as the enemy situation dictates.		
d. Check, if possible, positions from the enemy side to verify concealment, security and integration of sectors of fire.		
*9. The unit leader and subordinate leaders position themselves for control.		
*10. Unit leaders ensure their elements make final preparations. They take the following actions:		
a. Check all target acquisition and night observation devices.		
b. Maintain communications with the unit leader.		
c. Ensure the occupation and establishment of the positions are undetected by the enemy.		
d. Report to the unit leader when the positions are established.		
EXECUTE		
11. The unit maintains operations security (OPSEC), and executes active deception to deceive the enemy as to the ambush location and the strength of the unit.		
12. The security element detects the enemy. It takes the following actions:		
a. Alerts the unit leader.		
b. Reports the size of the target and the direction of movement.		
c. Reports any special weapons or equipment carried.		
*13. The unit leader alerts higher HQ and the remainder of the unit. He takes the following actions:		
a. Uses appropriate communication systems and or signals to notify the rest of the unit.		
b. Analyzes the situation; executes the ambush on command.		
*14. The unit leader initiates the ambush. He takes the following actions:		
a. Initiates the ambush with multiple, simultaneous antiarmor shots.		
b. Ensures unit/support elements deliver accurate fire.		
c. Ensures elements engage targets suitable to their weapons' characteristics.		

TASK STEPS AND PERFORMANCE MEASURES	GO	NO-GO
d. Uses the security element to block enemy attempts to maneuver and engage the ambush forces, to prevent reinforcement of the enemy in the EA, and to block enemy escaping from the EA.		
e. Employs indirect fires to support the ambush.		
f. Employs reserves, if designated, to block enemy escaping from the EA, to prevent enemy reinforcement in the EA, or to block counterattacks, if needed.		
g. Signals unit/support elements to lift or shift all supporting fires.		
h. Ensures compliance with ROE and that collateral damage is minimized.		
15. The unit secures enemy prisoners of war (EPW) as required.		
16. The unit prepares for withdrawal from the ambush site. It takes the following actions:		
a. Ensures all elements withdraw to subsequent positions or RP.		
b. Accounts for all personnel and equipment and report to higher HQ.		
*17. The unit leader directs or signals the withdrawal. He takes the following actions:		
a. Directs elements to withdraw to subsequent positions, RPs, according to OPORD, FRAGO, or SOPs while maintaining security.		
b. Uses planned fires to assist and cover the withdrawal or to complete destruction of the enemy in the EA, if needed.		
ASSESS		
*18. The unit leader reports to higher HQ.		
19. The unit consolidates and reorganizes as needed.		
20. The unit continues operations as directed.		
* indicates a leader task step		

SUPPORTING INDIVIDUAL TASKS

Task Number	Task Title
171-610-0001	Perform a Map Reconnaissance
061-283-6003	Adjust Indirect Fire
071-316-3006	Engage Targets with the TOW System on a BFV
071-054-0004	Engage Targets with an M136 Launcher
071-326-5770	Prepare a Platoon Sector Sketch

071-410-0010	Conduct a Leader's Reconnaissance
071-060-0005	Engage Targets with a Javelin
071-217-0022	Engage Targets (Tracker or Manual) with the Stryker Anti-Tank Guided Missile Vehicle Elevated Tow System
071-326-5503	Issue a Warning Order
071-331-0815	Practice Noise, Light, and Litter Discipline
071-326-5502	Issue a Fragmentary Order
071-326-5505	Issue an Operation Order at the Company, Platoon, or Squad Level

SUPPORTING COLLECTIVE TASKS

Task Number	Task Title
07-2-3027	Integrate Direct Fires (Platoon-Company)
07-2-3036	Integrate Indirect Fire Support (Platoon-Company)
07-2-5027	Conduct Consolidation and Reorganization (Platoon-Company)
07-2-5063	Conduct Composite Risk Management (Platoon-Company)
07-2-6063	Maintain Operations Security (Platoon-Company)
07-2-9006	Conduct a Passage of Lines as the Passing Unit (Platoon-Company)
07-2-9009	Conduct a Withdrawal (Platoon-Company)
07-3-1342	Occupy an Antiarmor Firing Position (Section-Platoon)
07-3-5045	Control TOW Fires (Section-Platoon)
07-3-9013	Conduct Action on Contact
07-3-9017	Conduct Actions at Danger Areas
19-3-3107	Process Detainee(s) at Point of Capture (POC)

SUPPORTING BATTLE/CREW DRILLS

Drill Number	Drill Title
07-3-D9501	React to Contact (Visual, IED, Direct Fire [includes RPG])
07-3-D9508	Establish Security at the Halt

STANDARDS TASK: Conduct Area Security (Platoon-Company) (07-2-1324)

(FM 3-21.10)   (FM 3-21.8)

CONDITIONS: The unit conducts operations as part of a higher headquarters (HQ) and receives an operation order (OPORD) or fragmentary order (FRAGO) to conduct area security operations at the location and time specified. All necessary personnel and equipment are available. Local populace and factions may or may not be cooperative. The unit has communications with higher, adjacent, and subordinate elements. The unit has been provided guidance on the rules of engagement (ROE). Coalition forces and noncombatants may be present in the operational environment. Some iterations of this task should be conducted during limited visibility conditions. This task should not be trained in mission-oriented protective posture 4 (MOPP4).

STANDARDS: The unit conducts area security according to the standing operating procedures (SOPs), the order, and/or higher commander's guidance. The unit establishes a force presence throughout the area of operations (AO). The unit prevents threat ground reconnaissance elements from directly observing friendly activities within the area being secured; and it prevents threat ground maneuver forces from penetrating the defensive perimeters established by the unit leader.

TASK STEPS AND PERFORMANCE MEASURES	GO	NO-GO
PLAN		
*1. Unit leaders gain and/or maintain situational understanding using available communications equipment, maps, intelligence summaries, situation reports (SITREPs), and other available information sources. Intelligence sources include company intelligence support team (CoIST), human intelligence (HUMINT), signal intelligence (SIGINT), and imagery intelligence (IMINT) to include unmanned aircraft systems (UASs) and unattended ground sensors (UGSs).		
*2. The unit leader receives an OPORD or FRAGO and issues a warning order (WARNO) to include at a minimum—		
a. The mission or nature of the area security.		
b. The time and place for issuing the OPORD.		
c. Units or elements participating in the area security.		
d. Specific tasks not addressed by unit SOPs.		

TASK STEPS AND PERFORMANCE MEASURES	GO	NO-GO
e. The timeline for the area security.		
*3. The unit leader plans for the mission using troop-leading procedures (TLPs). He takes the following actions:		
a. Conducts a map reconnaissance of the AO.		
(1) Identifies and marks boundaries for AO.		
(2) Identifies locations for possible observation posts and checkpoints.		
b. Determines liaison requirements.		
c. Coordinates for liaison officers, local guides, interpreters as required.		
d. Determines reporting requirements to higher HQ.		
e. Develops casualty evacuation (CASEVAC) procedures.		
f. Identifies security measures.		
g. Identifies areas where U.S. forces should not go (for example, religious shrines, areas where the peace mandate or other agreement restrict U.S. access).		
h. Identifies protection requirements.		
i. Determines resupply requirements.		
j. Plans for employment of augmentations to unit such as civil military detachments, military police teams, and sniper teams, as required.		
k. Develops task organization required to accomplish the mission.		
l. Addresses actions on chance contact with enemy.		
*4. The unit leader establishes a reserve force. He takes the following actions:		
a. Designates the reserve force element.		
b. Selects primary and alternate positions for the reserve force.		
c. Selects routes to projected places of employment.		
d. Designates control measures.		
e. Defines linkup procedures.		
f. Identifies conditions for employment.		
PREPARE		
*5. The unit leader provides intelligence requirements to security force.		
* 6. The unit leader identifies the security task required to be performed. He takes the following actions:		

TASK STEPS AND PERFORMANCE MEASURES	GO	NO-GO
a. Identifies the need for reconnaissance and/or combat patrols.		
b. Determines the need for checkpoints.		
c. Identifies the requirement for convoy escorts.		
d. Determines the need for observation posts (OPs).		
*7. The unit leader disseminates reports (if applicable), overlays, and other pertinent information to subordinates to keep them abreast of the situation.		
*8. The unit leader issues clear and concise tasking, orders and instructions to include ROE. He issues FRAGOs, as necessary, to address changes to the plan identified during the rehearsal.		
9. The unit conducts a rehearsal (includes rehearsal of reserve force).		
EXECUTE		
10. The unit establishes and occupies an outpost as required.		
11. The unit conducts area security mission. It takes the following actions:		
a. Executes patrols as required.		
(1) Conducts reconnaissance patrols when necessary. Takes the following actions:		
(a) Executes a route reconnaissance.		
(b) Executes an area reconnaissance.		
(c) Executes a zone reconnaissance.		
(d) Executes a point reconnaissance.		
(e) Executes a leader's reconnaissance.		
(2) Conducts combat patrols when needed. Takes the following actions:		
(a) Executes a raid patrol.		
(b) Executes an ambush patrol.		
(c) Executes a security patrol.		
(3) Establishes patrol routes and schedules as required.		
(4) Assigns mission to elements and supervises their activities.		
(5) Maintains communications with higher HQ and subordinate units.		
(6) Maintains capability to reinforce or support patrols with fires according to order, guidance, and or SOPs.		

TASK STEPS AND PERFORMANCE MEASURES	GO	NO-GO

NOTE: The unit's habitual use of elements to patrol selected areas should help the unit develop familiarity with the community and the area and build trust and confidence with the citizens. If cordon and search operations or vehicle inspections are required, units familiar with the area and the populace should conduct the mission.
      (7) Debriefs patrols as required.
   b. Establishes hasty or deliberate checkpoints. Takes the following actions:
      (1) Positions checkpoint in an area clear of hazards.
      (2) Positions checkpoint where it is visible.
      (3) Positions vehicles to deter resistance to Soldiers manning checkpoint.
      (4) Emplaces obstacles to slow traffic into search area.
      (5) Establishes a reserve.
      (6) Establishes a bypass lane.
      (7) Establishes communications within checkpoint area.
      (8) Designates search area.
      (9) Constructs and equips checkpoint.
   c. Secures routes.
   d. Conducts OP operations. Takes the following actions:
      (1) Identifies activities or locations to be observed.
      (2) Conducts reconnaissance to select OP sites across unit AO.
      (3) Assigns OP missions.
      (4) Repositions OPs as required.
      (5) Maintains capability to reinforce or support OP(s) by fires according to order, guidance, and/or SOPs.
   e. Executes convoy escorts for military or civilian movements as required.
   f. React to civil disturbances.
   g. Searches buildings. Takes the following actions:
      (1) Identifies object of search (for example, weapons, contraband, and so forth).		

TASK STEPS AND PERFORMANCE MEASURES	GO	NO-GO

(2) Ensures coordination has been conducted for required augmentation such as explosive ordinance disposal (EOD) or military working dogs (MWDs).
    (3) Reports inspection results according to ROE, higher HQ orders, or SOPs.
  h. Secures selected sites (for example, voting sites, refugee camps, schools, churches) according to ROE, and higher HQ orders. Takes the following actions:
    (1) Conducts reconnaissance to identify sites.
    (2) Assigns subordinate element missions.
  i. Enforces curfews. Takes the following actions:
    (1) Publicizes the curfew periods.
    (2) Monitors curfew compliance.
  j. Stabilizes areas with escalating tension. Takes the following actions:
    (1) Identifies potential "hot spots" of increased tension.
    (2) Determines which factions may be involved and their probable objectives.
    (3) Coordinates with factions to resolve real or perceived problems.
    (4) Dispatches coordination or liaison teams as required.
    (5) Reports developments of any de-stabilizing incidents and other changes to situation to higher HQ as required.
  k. Demonstrates resolve, confidence, commitment, and sensitivity for local customs and people living in the AO by attending local events.
  l. Coordinates for disposition of detained personnel, documents, equipment, and weapons.
  m. Commits the reserve force as required.
  n. Establishes an upgraded alert status for elements in affected and adjacent areas, as needed.
  o. Submits reports according to higher HQ order and SOPs.
ASSESS
12. The unit consolidates and reorganizes as needed.
13. The unit continues operations as directed.
* indicates a leader task step.		

SUPPORTING INDIVIDUAL TASKS

Task Number	Task Title
171-121-4045	Conduct Troop-Leading Procedures
171-610-0001	Perform a Map Reconnaissance
171-300-0008	Secure a Critical Area at Platoon Level
071-326-5502	Issue a Fragmentary Order
071-326-5503	Issue a Warning Order

SUPPORTING COLLECTIVE TASKS

Task Number	Task Title
07-2-1387	Employ a Reserve Force (Platoon-Company)
07-2-1405	Establish an Outpost (Platoon-Company)
07-2-1450	Secure Routes (Platoon-Company)
07-2-2054	Reconnoiter a Built-up Area (Platoon-Company)
07-2-4054	Secure Civilians During Operations (Platoon-Company)
07-2-5009	Conduct a Rehearsal (Platoon-Company)
07-2-5036	Conduct Coordination (Platoon-Company)
07-2-5045	Conduct Negotiations (Platoon-Company)
07-2-5063	Conduct Composite Risk Management (Platoon-Company)
07-2-6063	Maintain Operations Security (Platoon-Company)
07-2-9006	Conduct a Passage of Lines as the Passing Unit (Platoon-Company)
07-2-9051	Conduct a Cordon and Search (Platoon-Company)
07-3-9013	Conduct an Action on Contact
07-3-9016	Establish an Observation Post
07-3-9017	Conduct Actions at Danger Areas
07-3-9018	Enter and Clear a Building (Section-Platoon)
07-3-9023	Conduct a Presence Patrol
08-2-0004	Evacuate Casualties
19-3-2007	Conduct Convoy Security
19-3-2406	Conduct Roadblock and Checkpoint
19-3-4004	Conduct Civil Disturbance Control
44-3-3220	Perform Passive Air Defense Measures
44-3-3221	Perform Active Air Defense Measures

SUPPORTING BATTLE/CREW DRILLS

Drill Number	Drill Title
07-4-D9509	Enter and Clear a Room
05-3-D0016	Conduct the 5 Cs
05-3-D0017	React to an IED Attack While Mounted
05-3-D0015	React to an IED Attack While Dismounted

TASK: Conduct Support by Fire (Platoon-Company) (07-2-3000)

(FM 3-21.10)   (FM 3-21.8)

CONDITIONS: The unit conducts operations as part of a higher headquarters (HQ) and receives an operation order (OPORD) or fragmentary order (FRAGO) to conduct a support by fire (SBF). All necessary personnel and equipment are available. The unit has communications with higher, adjacent, and subordinate elements. The unit has been provided guidance on the rules of engagement (ROE). Coalition forces and noncombatants may be present in the operational environment. Some iterations of this task should be conducted during limited visibility conditions. Some iterations of this task should be performed in mission-oriented protective posture 4 (MOPP4).

STANDARDS: The unit conducts the SBF according to standing operating procedures (SOPs), the order, and/or higher commander's guidance. The unit occupies SBF positions undetected and suppresses or destroys enemy elements that could affect accomplishment of the supported force's mission. The unit maintains communications with the supported force.

TASK STEPS AND PERFORMANCE MEASURES	GO	NO-GO

PLAN
*1. Unit leaders gain and/or maintain situational understanding using available communications equipment, intelligence summaries, situation reports (SITREP), and other available information sources. Intelligence sources include company intelligence support team (CoIST), human intelligence (HUMINT), signal intelligence (SIGINT), and imagery intelligence(IMINT) to include unmanned aircraft systems (UASs) and unattended ground sensors (UGSs).
*2. The unit leader receives an OPORD or FRAGO and issues a warning order (WARNO) normally containing at a minimum—
    a.   The mission or nature of the operation.
    b.   The time and place for issuing the OPORD.
    c.   Units or elements participating in the operation.
    d.   Specific tasks not addressed by unit SOPs.
    e. The timeline for the operation.		

TASK STEPS AND PERFORMANCE MEASURES	GO	NO-GO

*3. The unit leader confirms friendly and enemy situations. He takes the following actions:
    a. Receives an updated report showing the location of forward and adjacent friendly elements, and the SBF.
    b. Receives an updated enemy situational template for added fratricide prevention and increased force protection, if applicable.
    c. Clarifies priority intelligence requirement (PIR) requirements.
    d. Confirms any changes to the higher HQ and unit task or purpose.
    e. Confirms any changes to the scheme of maneuver.
*4. The unit leader plans using troop-leading procedures (TLPs). He takes the following actions:
    a. Conducts analysis based on factors of mission, enemy, terrain and weather, troops and support available, time available, civil considerations (METT-TC).
    b. Considers the enemy's capabilities, likely courses of action (COAs), and specific weapons capabilities.
    c. Conducts a map reconnaissance. Takes the following actions:
        (1) Identifies tentative SBF positions (primary and alternate).
        (2) Identifies likely enemy avenues of approach (mounted and dismounted).
        (3) Identifies routes to and from SBF positions (primary and alternate).
        (4) Identifies tentative target reference points (TRPs).
        (5) Marks tentative dismount points on maps as appropriate.
    d. Plans the integration of direct and indirect fires according to higher HQ fire support plan.
    e. Conducts liaison with maneuver elements to integrate anti-fratricide measures.
    f. Plans primary and alternate triggers and/or signals for lifting or shifting direct and indirect fires.
    g. Plans forward observer (FO) positions so they can effectively adjust indirect fires.
    h. Develops control measures.		

TASK STEPS AND PERFORMANCE MEASURES	GO	NO-GO
i. Establishes sectors of observation and fire for each position. j. Establishes engagement priorities. k. Develops rules, signals, and methods of engagement. l. Develops criteria and signals for disengagement. m. Conducts coordination with maneuver force. He coordinates— 　(1) Control measures. 　(2) TRPs. 　(3) SBF position locations. n. Plans and coordinates sustainment. o. Plans to pre-position supplies, if necessary. p. Coordinates and synchronizes activities within each warfighting function. q. Organizes the unit as needed to accomplish the mission and/or compensate for combat losses. r. Designates the main effort and supporting effort. s. Addresses actions on chance contact with the enemy. PREPARE *5. The unit leader disseminates digital reports (if applicable), overlays, and other pertinent information to each element to keep them abreast of the situation. *6. The unit leader issues clear and concise tasking, orders and instructions to include ROE. 7. The unit conducts a rehearsal. *8. The unit leader issues a FRAGO, as needed, to address changes to the plan identified during the rehearsal. *9. The unit leader or designated representative and reconnaissance element conducts the reconnaissance based on factors of METT-TC. He takes the following actions: a. Selects SBF positions that— 　(1) Ensure the SBF element can place effective fires within the constraints of the terrain, on the enemy flanks, and provide overwatch within the primary weapon range. 　(2) Provide adequate cover and concealment. b. Secures SBF positions. c. Confirms and or selects TRPs.		

TASK STEPS AND PERFORMANCE MEASURES	GO	NO-GO

d. Identifies the avenues of approach for mounted and dismounted enemy elements.
    e. Updates the enemy situation.
    f. Leaves security element at SBF positions.
    g. Returns to the unit position.
*10. The unit leader adjusts the plan based on updated intelligence and reconnaissance effort.
*11. The unit leader updates the enemy situation.
*12. The unit leader disseminates updated digital reports (if applicable), overlays, and other pertinent information.
*13. The unit leader coordinates and/or synchronizes actions of all elements.
*14. The unit leader uses FRAGOs as needed to redirect actions of subordinate elements.
EXECUTE
 	
15. The unit moves tactically to and occupies designated SBF positions. It takes the following actions:
    a. Uses cover and concealed routes to prevent the enemy force from effectively engaging the SBF element.
    b. Occupies the most advantageous terrain that allows the placement of accurate fires on the enemy.
    c. Maintains local security.
    d. Uses natural or man-made obstacles on the position.
    e. Verifies fire procedures and control measures.
    f. Emplaces weapon systems. Takes the following actions:
       (1)   Emplaces weapon systems covering sectors of fire and observation and any other designated targets that increase flank shots on the enemy.
       (2)   Employs mortars to provide indirect fire support.
    g. Observes (continually) the maneuver force axis, route, sector, or direction of attack.
    h. Identifies known or suspected enemy positions that could engage the maneuver force.
*16. The unit leader positions self so he can view and control the battle.
17. The unit conducts overwatch as the situation dictates. It takes the following actions:		

TASK STEPS AND PERFORMANCE MEASURES	GO	NO-GO
a. Scans sectors of fire according to the SOPs or OPORD.		
b. Keeps maneuver force informed of the enemy situation and of any lapses in overwatch coverage.		
18. The unit employs direct fires. It takes the following actions:		
a. Acquires, suppresses, and/or destroys identified enemy elements using the appropriate weapon systems.		
b. Prevents the enemy from placing accurate fires against the protected force. Takes the following actions:		
(1) Maintains security to prevent the enemy from engaging the maneuver force.		
(2) Repositions as needed to maintain effective observation and/or fires on the enemy or to prevent the enemy from acquiring or destroying the maneuver force.		
(3) Employs dismounted Soldiers, if necessary.		
c. Lifts or shifts fires on order or by predetermined signal.		
d. Ceases fire on order or by predetermined signal.		
*19. The unit leader employs indirect fires to suppress, obscure, or destroy the enemy or to screen the movement of the maneuver force.		
ASSESS		
20. The unit fixes, suppresses, or destroys the enemy according to the OPORD.		
21. The unit consolidates and reorganizes as needed.		
22. The unit continues operations as directed.		
* Indicates a leader task step.		

SUPPORTING INDIVIDUAL TASKS

Task Number	Task Title
171-610-0001	Perform a Map Reconnaissance
061-283-6003	Adjust Indirect Fire
071-410-0010	Conduct a Leaders Reconnaissance
071-326-5503	Issue a Warning Order
071-600-0009	Coordinate with Supported Units
171-121-4045	Conduct Troop leading Procedures
071-326-5502	Issue a Fragmentary Order
071-030-0004	Engage Targets with an MK 19 Grenade Machine Gun

071-054-0004	Engage Targets with an M136 Launcher
071-314-0012	Engage Targets with the 25-mm Automatic Gun on a BFV
071-316-3006	Engage Targets with the TOW System on a BFV

SUPPORTING COLLECTIVE TASKS

Task Number	Task Title
07-2-1324	Conduct Tactical Movement (Platoon-Company)
07-2-1396	Employ Obstacles (Platoon-Company)
07-2-3027	Integrate Direct Fires (Platoon-Company)
07-2-3036	Integrate Indirect Fire Support (Platoon-Company)
07-2-5009	Conduct a Rehearsal (Platoon-Company)
07-2-5027	Conduct Consolidation and Reorganization (Platoon-Company)
07-2-5063	Conduct Composite Risk Management (Platoon-Company)
07-2-6063	Maintain Operations Security (Platoon-Company)
08-2-0003	Treat Casualties
08-2-0004	Evacuate Casualties

SUPPORTING BATTLE/CREW DRILLS

Drill Number	Drill Title
07-3-D9501	React to Contact (Visual, IED, Direct Fire [includes RPG])
07-3-D9504	React to Indirect Fire

TASK: Integrate Indirect Fire Support (Platoon-Company) (07-2-3036)

(FM 3-21.10)   (FM 3-21.8)

CONDITIONS: The unit conducts operations as part of a higher headquarters (HQ) and integrates fire support for the mission. All necessary personnel and equipment are available. The unit has the battalion (BN) target list; communicates with higher, adjacent, and subordinate elements; and has guidance on the rules of engagement (ROE). Coalition forces and noncombatants may be present in the operational environment. Some iterations of this task should be conducted during limited visibility conditions and performed in mission-oriented protective posture 4 (MOPP4).

STANDARDS: The unit plans and integrates fire support according to standing operating procedures (SOPs), the order, and/or higher commander's guidance. Unit leaders and/or the fire support team (FIST) determine the desired effect fires should have on the enemy. Unit leaders and/or the FIST plan, integrate, and coordinate indirect fires to support all phases of the operation. Unit leaders and/or forward observers (FOs) employ indirect fires using the correct "call for fire" format and procedures.

TASK STEPS AND PERFORMANCE MEASURES	GO	NO-GO

PLAN
*1. Unit leaders gain and/or maintain situational understanding using available communications equipment, maps, intelligence summaries, situation reports (SITREPs), and other available information sources. Intelligence sources include company intelligence support team (CoIST), human intelligence (HUMINT), signal intelligence (SIGINT), and imagery intelligence (IMINT) to include unmanned aircraft systems (UASs) and unattended ground sensors (UGSs).
*2. Unit leaders confirm friendly and enemy situations. They take the following actions:
    a.   Receive an updated report showing the location of forward and adjacent friendly elements, if applicable.
    b.   Receive an updated enemy situational template for added fratricide prevention and increased force protection, if applicable.
    c.   Clarify priority intelligence requirements (PIRs).
    d.   Confirm changes to the higher HQ and unit task or purpose.
    e. Confirm changes to the scheme of maneuver.		

TASK STEPS AND PERFORMANCE MEASURES	GO	NO-GO

*3. Unit leaders perform a map reconnaissance. They take the following actions:
    a. Identify tentative target reference points (TRPs).
    b. Identify probable or known enemy locations.
*4. Unit leaders and the FIST plan fire support. They take the following actions:
    a. Determine desired effect on the enemy (suppress, isolate, obscure, neutralize, destroy, deceive, or disrupt).
    b. Plan priority of fires (should support the main effort).
    c. Identify priority targets.
    d. Plan close air support.
    e. Identify ammunition restrictions and controlled supply rate.
    f. Develop graphical fire control measures to include measures to initiate, lift, or shift fires.
    g. Confirm whether the use of smoke, scatterable mines, illumination, or dual purpose improved conventional munitions is restricted and who controls them.
    h. Determine communications procedures to use when calling for fire.
    i. Determine when and under what circumstances to engage targets.
    j. Determine the method of engagement and method of control to be used.
    k. Develop the indirect fire plan at the same time as the offensive scheme of maneuver. Take the following actions:
        (1) Integrate direct fires and indirect fires to support maneuver throughout the operation.
        (2) Plan fires that support the commander's intent and scheme of maneuver.
        (3) Plan fires to support all phases of the attack.
        (4) Plan fires for targets of concern (targets that may deter the success of the maneuver).
        (5) Plan smoke to screen the unit when crossing a danger area, breaching an obstacle, or to obscure known or suspected enemy positions.		

TASK STEPS AND PERFORMANCE MEASURES	GO	NO-GO

1.   Develop the indirect fire plan to support the defensive scheme of maneuver. Take the following actions:
        (1) Plan fires that support the commander's intent.
        (2) Plan fires on all likely enemy positions and on areas the enemy may use in the attack, such as:
                (a) Observation posts (OPs).
                (b) Support positions.
                (c) Avenues of approach.
                (d) Assault positions.
                (e) Dead space.
                (f) Flanks.
                (g) Defiles.
        (3) Plan fires in front of, on top of, and behind friendly positions to stop likely penetrations or to support a counterattack.
        (4) Integrate final protective fires (FPFs) into the unit fire and obstacle plans.
        (5) Plan fires that cover planned or existing obstacles.
        (6) Plan smoke to screen friendly movements. (Defending units should use smoke sparingly.)
        (7) Plan illumination.
NOTE: Unit leaders normally retain control of illumination in the defense.
*5. Unit leaders and the FIST, if available, prepare for fire support execution. They take the following actions:
        a.   Prepare an observation plan. Take the following actions:
                (1) Designate primary and alternate observers.
                (2) Brief observers on target tasks and purposes.
                (3) Identify engagement area (EA).
                (4) Ensure observers are positioned to observe EA.
                (5) Consider available assets such as the laser range finders.
        b. Prepare a trigger to initiate fires for each target. Take the following actions:
                (1) Include engagement criteria.
                (2) Prepare trigger lines based on the following:		

TASK STEPS AND PERFORMANCE MEASURES	GO	NO-GO

(a) Rate of travel by enemy forces to the engagement area.
          (b) Amount of time required to call for fires.
          (c) Time of flight of the indirect fire rounds.
          (d) Clearance of fires at the unit and element level.
          (e) Possible adjustment times.
    c.   Prepare a trigger for lifting or shifting fires. (For offensive operations, use a minimum safe line.)
PREPARE
6.   The unit conducts a rehearsal. It takes the following actions:
    a.   Involves observers in unit rehearsals.
    b.   Ensures the unit's primary and alternate communications systems supports the fire support plan.
    c.   Ensures precombat checks have been conducted on equipment according to the SOPs.
*7. Unit leaders or designated representatives conduct a reconnaissance to confirm the indirect fire plan, if possible.
*8. Unit leaders distribute the indirect fire support plan/execution matrix to subordinate leaders as a part of the OPORD. They take the following actions:
    a.   Provide a copy of the fire plan to higher HQ.
    b.   Incorporate the fire support plan into rehearsals.
*9. Unit leaders use FRAGOs as necessary to redirect actions of subordinate elements.
EXECUTE
*10. Unit leaders or the FO employ indirect fire support using available communications. They take the following actions:
    a. Ensure all available supporting fires are executed in a timely manner and accomplish the prescribed result according to the fire plan and execution matrix.
    b. Ensure the employment of smoke does not degrade the unit mission.
    c. Adjust the priorities of fire as the battle progresses.
    d. Use the combat observation lasing team (COLT) when available.
    e. Call for fire, including—		

TASK STEPS AND PERFORMANCE MEASURES	GO	NO-GO
(1) Proper standardized call-for-fire (CFF) format.		
(2) Proper radio communications procedures to call for fire.		
(3) Observer identification and warning order (adjust fire, fire for effect, suppress, or immediate suppression).		
(4) Target location methods (grid, polar, or shift from a known point).		
(5) Target description using size and/or shape, nature/nomenclature, activity, and protection/posture.		
(6) Various techniques for area adjustment, such as—		
(a) Successive bracketing.		
(b) One round adjustment.		
(c) Creeping fire.		
(7) Correct observer target factor and angular deviation.		
(8) Subsequent corrections.		
(9) Fire for effect when burst is within 50 meters.		
*11. Unit leaders or the FO conduct "fire for effect missions." They take the following actions:		
a. Ensure the impact of the adjustment rounds is close enough to have the desired effects with the first volley fired.		
b. Request appropriate shell/fuse combination.		
NOTE: If desired effects are not achieved, the observer adjusts the rounds and repeats, changes shell/fuse combination, and requests additional fire for effect volleys.		
ASSESS		
*12. Unit leaders direct the observer to take the following actions if the desired effects are not achieved. The observer—		
a. Adjusts the rounds and repeats.		
b. Changes the shell and or fuse combination.		
c. Requests additional fire for effect volleys.		
*13. Unit leaders or the FO conduct immediate suppression missions. They take the following actions:		
a. Identify the target (observer).		
b. Plot the target accurately.		
c. Transmit complete call for fire.		

TASK STEPS AND PERFORMANCE MEASURES	GO	NO-GO

d. Ensure an accurate target location is close enough to have the desired effects with the first volley fired.
    e. Ensure final suppression rounds are within 150 meters of the target (if necessary).
NOTE: If desired effects are not achieved, the observer adjusts the rounds and repeats, changes shell/fuse combination, and requests additional fire for effect volleys.
14. The FO observes munitions effects and reports battle damage assessments. He takes the following actions:
    a. Estimates the extent of damage to the target and or casualties.
    b. Reports damage assessment to the fire direction center (FDC) providing fires.
15. The FO, with laser locator, conducts a high burst and or mean point of impact registration (field artillery [FA] only), when directed. (The FDC transmits orienting data to observer.) He takes the following actions:
    a. Orients using orienting data.
    b. Announces to FDC ready to observe.
    c. Lases the burst.
    d. Records and transmits burst location to FDC until FDC terminates registration.
*16. Unit leaders or the FO register, confirm, and adjust a parallel sheaf for mortars. They take the following actions:
    a. Use successive bracketing.
    b. Send appropriate corrections to FDC.
    c. Adjust sheaf to within a 50-meter range and a 40-meter lateral spread between rounds.
* indicates a leader task step.		

SUPPORTING INDIVIDUAL TASKS:

Task Number	Task Title
171-610-0001	Perform a Map Reconnaissance
061-284-3040	Engage Targets with Close Air Support
061-283-6003	Adjust indirect Fire

SUPPORTING COLLECTIVE TASKS

Task Number	Task Title
07-2-3027	Integrate Direct Fires (Platoon-Company)
07-2-5009	Conduct a Rehearsal (Platoon-Company)

| 07-2-5063 | Conduct Composite Risk Management (Platoon-Company) |
| 07-2-6063 | Maintain Operations Security (Platoon-Company) |

SUPPORTING BATTLE/CREW DRILLS

Drill Number	Drill Title
07-3-D9406	Knock Out Bunker
17-3-D8008	React to an Obstacle

TASK: Secure Civilians During Operations (Platoon-Company) (07-2-4054)

(FM 3-21.10)   (FM 3-21.8)

CONDITIONS: The unit conducts operations as part of a higher headquarters (HQ) and receives an operation order (OPORD) or fragmentary order (FRAGO) to secure civilians to protect them from injury due to combat. Some may be refugees and others may be inhabitants of the area in which the unit operates. Some may be openly hostile. All necessary personnel and equipment are available. The unit communicates with higher, adjacent, and subordinate elements. The unit has guidance on the rules of engagement (ROE). Coalition forces and noncombatants may be present in the operational environment. Some iterations of this task should be conducted during limited visibility conditions and performed in mission-oriented protective posture 4 (MOPP4).

STANDARDS: The unit secures civilians during operations according to the standing operating procedures (SOPs), the order, and/or the higher commander's guidance. The unit identifies and segregates combatants and noncombatants, and searches, safeguards, and moves them out of the immediate area of operations.

TASK STEPS AND PERFORMANCE MEASURES	GO	NO-GO
PLAN *1. Unit leaders gain and/or maintain situational understanding using available communications equipment, maps, intelligence summaries, situation reports (SITREPs), and other available information sources. Intelligence sources include company intelligence support team (CoIST), human intelligence (HUMINT), signal intelligence (SIGINT), and imagery intelligence (IMINT) to include unmanned aircraft systems (UASs) and unattended ground sensors (UGSs). *2. The unit leader receives an OPORD or a fragmentary order FRAGO directing unit to secure civilians. Unit leader issues a warning order (WARNO) to element leaders ensuring that subordinate leaders have sufficient time for their own planning and preparation needs. The WARNO must include— a. Tentative unit organization for the securing of civilians.		

TASK STEPS AND PERFORMANCE MEASURES	GO	NO-GO
b.Location and tentative timeline of the operation, including movement times and no later than times for execution.		
c. Guidance directing the unit to conduct rehearsals; initiate movement; conduct reconnaissance tasks, and the commander's critical information requirement (CCIR).		
*3. Unit leaders conduct troop-leading procedures with emphasis on the following:		
a. Interpreters to help interface with the local populace, if necessary.		
b.Control measures for expected or unexpected situations.		
c. Organization of the unit to accomplish the mission and/or compensate for combat losses.		
PREPARE		
*4. Unit leaders disseminate reports and overlays to each subordinate element to keep them abreast of the situation.		
*5. Unit leaders issue clear and concise tasking, orders, and instructions to include ROE.		
6. The unit conducts a rehearsal.		
EXECUTE		
7. Unit leaders or designated representatives supervise the operation. They take the following actions:		
a. Ensure civilians are treated with respect.		
b.Ensure elements understand the ROE.		
c. Ensure elements/soldiers understand procedures for dealing with news media.		
d.Use FRAGOs as necessary to redirect actions of subordinate elements.		
8. Designated elements secure civilians. They take the following actions:		
a. Maintain 360-degree and three-dimensional security in the AO in which civilians are gathered.		
b.Segregate civilians identified as being combatants or suspected war criminals and treat them like enemy prisoners of war (EPWs).		
c. Report the situation and status in a timely manner to higher HQ.		
d.Assign personnel to search civilians. (Keep identification papers with civilians under all circumstances, regardless of status.)		

TASK STEPS AND PERFORMANCE MEASURES	GO	NO-GO
e. Restrain and detain noncombatants who do not follow instructions, including—		
(1) Safeguard noncombatants and provide humane but firm treatment at all times.		
(2) Move noncombatants away from the immediate combat area and safeguard against hostile fire.		
f. Provide food, water, and medical attention based upon the medical ROE for civilian medical treatment.		
g. Assign guards to escort the civilians, including—		
(1) Evacuate civilians to a processing and or reception station or to an intermediate collection point run by higher HQ.		
(2) Ensure that guards escorting the civilians are prepared to give concise information to the processing/reception station or intermediate collection point about the original location of the civilians and their actions since being encountered (for example, reluctant, totally uncooperative, hostile).		
9. The unit gives proper consideration to the situation of the press and local officials.		
ASSESS		
10. The unit follows ROE guidance as to whether the local civilians and officials are to be considered friendly, hostile, or uncertain.		
11. The unit continues operations as directed.		
* indicates a leader task step.		

SUPPORTING INDIVIDUAL TASKS

Task Number	Task Title
171-121-4045	Conduct Troop-Leading Procedures
191-377-4254	Search a Detainee
191-377-4256	Guard Detainees

SUPPORTING COLLECTIVE TASKS

Task Number	Task Title
07-2-5009	Conduct a Rehearsal (Platoon-Company)
07-2-5063	Conduct Composite Risk Management (Platoon-Company)
07-2-6063	Maintain Operations Security (Platoon-Company)
08-2-0003	Treat Casualties
08-2-0004	Evacuate Casualties

19-3-3107   Process Detainee(s) at Point of Capture (POC)
07-2-5081   Conduct Troop-leading Procedures (Platoon-Company)

SUPPORTING BATTLE/CREW DRILLS
Drill Number   Drill Title
05-3-D0016   Conduct the 5 Cs

TASK: Conduct Consolidation and Reorganization (Platoon-Company) (07-2-5027)

(FM 3-21.10)   (FM 3-21.8)

CONDITIONS: The unit conducts operations as part of a higher headquarters (HQ) and is in contact with the enemy. The unit must consolidate and reorganize. The unit communicates with higher, adjacent, and subordinate elements. Enemy forces have withdrawn to hasty defensive positions but have the capability to counterattack. The unit has guidance on the rules of engagement (ROE). Coalition forces and noncombatants may be present in the operational environment. Some iterations of this task should be performed in mission-oriented protected posture 4 (MOPP 4).

STANDARDS: The unit consolidates and reorganizes according to the standing operating procedures (SOP) and/or higher commander's guidance. The unit occupies a hasty fighting position with sectors of fire, establishes security, accounts for all personnel and equipment, and reestablishes the chain of command. Wounded in action (WIAs) are identified, stabilized, and prepared for evacuation. Killed in action (KIAs) are identified and prepared for evacuation. Ammunition and supplies are redistributed as needed.

TASK STEPS AND PERFORMANCE MEASURES	GO	NO-GO

PLAN
*1. Unit leaders gain and/or maintain situational understanding using available communications equipment, maps, intelligence summaries, situation reports (SITREPs), and other available information sources. Intelligence sources include company intelligence support team (CoIST), human intelligence (HUMINT), signal intelligence (SIGINT), and imagery intelligence (IMINT) to include unmanned aircraft systems (UASs) and unattended ground sensors (UGSs).
*2. Unit leaders confirm friendly and enemy situations. They receive an updated—
    a. Report showing the location of forward and adjacent friendly elements.
    b. Enemy situational template for added fratricide prevention and increased force protection.
*3. Unit leaders conduct troop-leading procedures.
PREPARE		

TASK STEPS AND PERFORMANCE MEASURES	GO	NO-GO

*4. Unit leaders position the observation post (OP) forward to provide security. They ensure that—
    a. Members are alert for a possible counterattack.
    b. The unit main body is not engaged without warning.
 EXECUTE
 5.   The unit occupies hasty fighting positions near the objective. It takes the following actions:
    a. Establishes local security, including—
      (1) Evaluates terrain thoroughly.
      (2) Positions the elements using the clock or the terrain feature technique.
      (3) Mans key weapons, as required by factors of mission, enemy, terrain and weather, troops and support available, time available, civil considerations (METT-TC).
    b. Destroys all organized resistance.
    c. Conducts reconnaissance of objective and/or area of operations (AO) to ensure it is free of enemy.
    d. Defends against enemy counterattacks.
    e. Begins decontamination operations, if required and as factors of METT-TC dictate.
    f. Establishes the chain of command.
    g. Establishes communications.
 *6. Unit leaders assign elements temporary sectors of fire.
 *7. Unit leaders ensure subordinate leaders adjust positions to cover likely avenues of approach and ensure mutual support between elements and adjacent units.
 *8. The unit secures enemy prisoners of war (EPWs).
 *9. Unit leaders report intelligence information of immediate value to next higher HQ.
 *10. Unit leaders supervise redistribution of ammunition and equipment.
 *11. Unit leaders provide ammunition, casualty, and equipment (ACE) reports to the headquarters.
 *12. Unit leaders coordinate resupply.
 *13. The unit treats and evacuates casualties.
 *14. The unit processes captured documents and/or equipment as required.
 ASSESS		

TASK STEPS AND PERFORMANCE MEASURES	GO	NO-GO
*15. The unit continues operations as directed. *indicates a leader task step		

SUPPORTING INDIVIDUAL TASKS

Task Number	Task Title
171-121-4045	Conduct Troop-Leading Procedures
171-121-4038	Supervise Local Security
031-507-3014	Supervise Decontamination Procedures
113-571-1022	Perform Voice Communications
081-831-1058	Supervise Casualty Treatment and Evacuation
071-940-0002	Conduct Resupply of a Platoon
301-371-1200	Process Captured Materiel

SUPPORTING COLLECTIVE TASKS

Task Number	Task Title
07-2-6063	Maintain Operations Security (Platoon-Company)
07-3-9016	Establish an Observation Post
08-2-0003	Treat Casualties
08-2-0004	Evacuate Casualties
19-3-3107	Process Detainee(s) at Point of Capture (POC)

SUPPORTING BATTLE/CREW DRILLS

Drill Number	Drill Title
05-3-D0016	Conduct the 5 Cs
07-3-D9507	Evacuate a Casualty (Dismounted and Mounted)

TASK: Conduct an Attack (Platoon-Company) (07-2-9001)

(FM 3-21.10)   (FM 3-21.8)

CONDITIONS: The unit conducts operations independently or as part of a higher headquarters (HQ) and receives an operation order (OPORD) or fragmentary order (FRAGO) to assault an objective at the location and time specified. The unit is located in an assembly area and provides its own security. All necessary personnel and equipment are available. Indirect fire and close air support (CAS) are available. The unit communicates with higher, adjacent, and subordinate elements. The unit has guidance on the rules of engagement (ROE) and may also have specific mission instructions, such as a peace mandate, terms of reference, and a status-of-forces agreement (SOFA). Military and civilian, joint and multinational partners, and news media may be present in the operational environment (OE). Some iterations of this task should be performed in mission-oriented protective posture 4 (MOPP4).

STANDARDS: The unit conducts the attack in according to the unit standing operating procedures (SOPs), the order, and or higher commander's guidance. The unit moves tactically from the assembly area to the line of departure (LD) and then to assault, support, or breach positions using the appropriate formation and technique. The unit provides supporting fires. The unit suppresses enemy forces on or near the objective, assaults the objective to destroy or capture, or forces the enemy to withdraw. The unit complies with the ROE, mission instructions, higher HQ orders, and other special orders.

TASK STEPS AND PERFORMANCE MEASURES	GO	NO-GO

PLAN
*1. Unit leaders receive an OPORD or a FRAGO that requires their unit to conduct an attack and issue a warning order (WARNO) to the unit according to troop-leading procedures (TLPs) and unit SOPs. (Refer to Task 07-2-5081 Conduct Troop-leading Procedures [Platoon-Company].) The WARNO must include—
  a. Tentative unit organization for the attack, identifying the security forces, main body, reserve, and sustaining organization, as applicable.
  b. Location and tentative timeline for the attack, including movement times and no later than time for execution.		

TASK STEPS AND PERFORMANCE MEASURES	GO	NO-GO
c. Guidance directing the unit to conduct rehearsals; initiate movement and reconnaissance tasks, and ensure the commander's critical information requirements (CCIRs).		
d. Instructions to obtain markers and special breach equipment, if needed, and other special equipment or additional supplies required for the attack.		
*2. Unit leaders develop a tentative plan according to the TLPs. They take the following actions:		
a. Conduct mission analysis, using company intelligence support team (COIST), focusing on the mission given, enemy forces and their capabilities, terrain and weather effects, troops and time available to execute the operation, and civil considerations. They should take advantage of maps, imagery, unmanned aircraft systems (UASs), unattended ground sensors (UGSs), and other available capabilities.		
b. Develop a tentative course of action by taking the following actions:		
(1) Identify a tentative objective rally point (ORP), if necessary, and a tentative assault position.		
(2) Identify tentative security, support by fire, and assault positions.		
(3) Identify routes to and from the ORP, if used, and objective.		
(4) Mark tentative dismount and/or remount points on maps as appropriate.		
(5) Plan, integrate, and coordinate direct fire support, indirect fire support, and/or CAS to achieve one or more operational goals during each phase of the operation. These goals comprise plans to—		
(a) Suppress enemy antitank or other weapon systems that inhibit movement.		
(b) Fix or neutralize bypassed enemy elements.		
(c) Prepare enemy positions for an assault.		
(d) Obscure enemy observation or screen friendly maneuver.		
(e) Support breaching operations.		
(f) Illuminate enemy positions.		

TASK STEPS AND PERFORMANCE MEASURES	GO	NO-GO
(g) Employ available weapons systems (tanks, antiarmor) according to doctrine.		
(6) Plan and coordinate sustainment activities to assist maneuver elements in maintaining the momentum of the attack, including plans for—		
(a) The increased consumption of Class III and Class V supplies.		
(b) Casualty evacuation.		
(c) Increased equipment maintenance requirements.		
(d) Positioning sustainment assets as far forward as possible.		
(7) Develop control measures for movement to the objective and fire support throughout the operation.		
(8) Develop contingency plans for actions on chance contact with the enemy crossing of LDAs identified during mission analysis.		
(9) Conduct composite risk management to identify, assess, develop, and implement controls for hazards and to mitigate associated risks. (Refer to Task 07-2-5063, Conduct Composite Risk Management [Platoon-Company].)		
(10) Task-organize the unit into a support, assault, breach element, and security force accounting for special tasks such as quartering parties and reconnaissance and surveillance (R&S) teams.		
3. The unit begins necessary movement to meet all required timelines indicated in the OPORD.		
*4. Unit leaders conduct a leader's reconnaissance. They take the following actions:		
a. Pinpoint the objective.		
b. Establish security at the objective.		
c. Determine the enemy's size, location, disposition, and most probable course of action on the objective.		
d. Determine where the enemy is most vulnerable to attack and where the support element can best place fires on the objective.		
e. Verify and update intelligence information.		
f. Determine whether to conduct the assault mounted or dismounted, if applicable.		
g. Select security, support, and assault positions.		

TASK STEPS AND PERFORMANCE MEASURES	GO	NO-GO
h. Leave a surveillance team to observe the objective.		
i. Return to the unit position.		
*5. Unit leaders adjust the plan based on updated intelligence and reconnaissance efforts.		
*6. Unit leaders issue the OPORD and use FRAGOs, as necessary, to redirect actions of subordinate elements.		
PREPARE		
7. The unit prepares for attack by taking the following actions:		
a. Conducting a rehearsal.		
b. Completing final inspections.		
*8. Unit leaders supervise subordinate TLPs to ensure planning and preparations are on track and consistent with the unit commander's intent.		
9. The unit issues FRAGOs as needed to address changes to the plan identified during the rehearsal.		
EXECUTE		
10. The unit executes the attack. It takes the following actions:		
a. Moves to the line of departure (LD) using a technique and formation based on the factors of METT-TC (may be executed by other unit leaders while the unit leader is forward conducting a leader's reconnaissance).		
b. Navigates from checkpoint to checkpoint or phase line by using basic land navigation skills supplemented by precision navigation.		
c. Moves from the LD through the assault position to support positions, assault positions, or breach or bypass sites. Pauses in the assault position, if absolutely necessary, to ensure synchronization of all friendly forces. Takes the following actions:		
(1) Moves using the designated and/or appropriate movement technique.		
(2) Uses cover and concealment.		
(3) Communicates primarily by FM radio and signals (embedded digital reports if applicable) during movement.		
(4) Uses smoke and supporting fire if detected.		
(5) Executes contingency plans developed by the unit leader, if needed.		

TASK STEPS AND PERFORMANCE MEASURES	GO	NO-GO
d. Conducts the assault mounted. (A mounted assault is only conducted against light resistance or when there are no heavy antiarmor weapons on the objective.) Takes the following actions:		
(1) Does not stop after moving forward of the assault position.		
(2) Controls supporting fires to support risk management initiatives.		
(3) Isolates the objective, which includes—		
(a) Preventing the enemy from reinforcing the objective.		
(b) Placing suppressive fires on the most dangerous enemy positions.		
(c) Maintaining visual observation of suppressive fires just forward of the breach and assault elements.		
(d) Positioning or repositioning security elements and weapons systems to provide continual suppressive fire to aid the actions of the assault element as it moves across the objective.		
(e) Using FM radio or predetermined visual signals to communicate with the breach and assault elements.		
(4) Conducts initial breach of obstacles, if required.		
(5) Assaults the objective, which includes—		
(a) Leading with tanks, if available, and uses armed vehicles to provide supporting fires while moving.		
(b) Moving onto the objective by conducting fire and movement.		
(c) Dismounting if the enemy begins to place effective antiarmor fires on assaulting element. (Vehicles move to support positions.)		
(d) Using indirect fires to isolate portions of the objective area to obscure enemy element and/or to screen the movement of the assault element.		
(e) Ensuring bypassed enemy cannot place effective fires on tanks and or armed vehicles.		

TASK STEPS AND PERFORMANCE MEASURES	GO	NO-GO
(f) Destroying enemy forces, captures enemy forces, and/or forces their withdrawal from the objective area according to the unit leader's intent.		
e. Conducts the assault dismounted. Takes the following actions:		
(1) Does not stop after moving forward of the assault position.		
(2) Controls supporting fires to support risk management initiatives.		
f. Isolates and suppresses the enemy on the objective. Takes the following actions:		
(1) Prevents the enemy from reinforcing or leaving from the objective.		
(2) Places suppressive fires on the most dangerous enemy positions.		
(3) Shifts suppressive fires to allow the breach element to penetrate the objective.		
(4) Maintains visual observation of suppressive fires just forward of the breach and assault elements.		
(5) Positions or repositions weapons systems to provide continual suppressive fire to aid the actions of the assault element as it moves across the objective (done by unit leader or designated representative).		
(6) Uses FM radio or predetermined visual signals to communicate with the breach and assault element or both.		
g. Conducts initial breach of obstacles, if required.		
h. Assaults the objective. Takes the following actions:		
(1) Uses armed vehicles to provide supporting fires, if available.		
(2) Moves onto the objective by conducting fire and movement.		
(3) Uses indirect fires to isolate portions of the objective area to obscure enemy element and/or to screen the movement of the assault element.		
(4) Ensures bypassed enemy cannot place effective fires on unit elements.		

TASK STEPS AND PERFORMANCE MEASURES	GO	NO-GO
(5) Destroys enemy forces, captures enemy forces, and/or forces their withdrawal from the objective area according to the unit leader's intent.		
i. Occupies defensible positions as needed. Takes the following actions:		
(1) Assaults through the objective to occupy defensible terrain beyond the objective.		
(2) Prepares for a counterattack.		
11. The unit conducts consolidation and reorganization. (Refer to Task 07-2-5027 Conduct Consolidation and Reorganization [Platoon-Company].) It takes the following actions:		
a. Establishes security on the objective.		
b. Conducts reconnaissance of area.		
c. Reorganizes elements and mans keys weapons to compensate for combat losses.		
d. Redistributes ammo, supplies, and equipment as needed.		
e. Secures, processes, and evacuates enemy prisoners of war and/or other detainees according to unit SOPs and METT-TC.		
f. Treats and evacuates casualties.		
g. Processes captured documents and or equipment as required.		
h. Reports SITREP to higher using Force XXI Battle Command Brigade and Below (FBCB2), FM, or other tactical means.		
ASSESS		
12. The unit continues operations as directed.		
* indicates a leader task step.		

SUPPORTING INDIVIDUAL TASKS

Task Number	Task Title
171-121-4045	Conduct Troop Leading Procedures
171-610-0001	Perform a Map Reconnaissance
071-326-5630	Conduct Movement Techniques by a Platoon
171-121-3009	Control Techniques of Movement
061-283-6003	Adjust Indirect Fire
071-025-0007	Engage Targets with an M240B Machine Gun
071-010-0006	Engage Targets with an M249 Machine Gun
301-348-1050	Report Information of Potential Intelligence Value

SUPPORTING COLLECTIVE TASKS

Task Number	Task Title
07-2-1342	Conduct Tactical Movement (Platoon-Company)
07-2-3000	Conduct Support by Fire (Platoon-Company)
07-2-3027	Integrate Direct Fires (Platoon-Company)
07-2-3036	Integrate Indirect Fire Support (Platoon-Company)
07-2-4054	Secure Civilians During Operations (Platoon-Company)
07-2-5009	Conduct a Rehearsal (Platoon-Company)
07-2-5027	Conduct Consolidation and Reorganization (Platoon-Company)
07-2-6045	Employ Camouflage, Concealment, and Deception Techniques (Platoon-Company)
07-2-9006	Conduct a Passage of Lines as the Passing Unit (Platoon-Company)
07-3-9013	Conduct Action on Contact
07-3-9017	Conduct Actions at Danger Areas
08-2-0003	Treat Casualties
08-2-0004	Evacuate Casualties
44-3-3220	Perform Passive Air Defense Measures
44-3-3221	Perform Active Air Defense Measures

SUPPORTING BATTLE/CREW DRILLS

Drill Number	Drill Title
07-3-D9501	React to Contact (Visual, IED, Direct Fire [includes RPG])
07-3-D9410	Enter a Trench to Secure a Foothold
07-3-D9412	Breach of a Mined Wire Obstacle

TASK: Conduct a Bypass (Platoon-Company) (07-2-9002)

(FM 3-21.10)   (FM 3-21.8)

CONDITIONS: The unit conducts operations as part of a higher headquarters (HQ) and receives an operation order (OPORD) or fragmentary order (FRAGO) to bypass an obstacle, position, or enemy force to maintain the momentum of advance at the location and time specified. The unit is ordered to avoid becoming decisively engaged. All necessary personnel, equipment, indirect fire, and close air support (CAS) are available. The unit has communications with higher, adjacent, and subordinate elements. The unit has received guidance on the rules of engagement (ROE) and may also have specific mission instructions, such as a peace mandate, terms of reference, and a status of forces agreement (SOFA). Military and civilian, joint and multinational partners, and news media may be present in the operational environment (OE). Some iterations of this task should be performed in mission-oriented protective posture 4 (MOPP4).

STANDARDS: The unit conducts the bypass according to the unit standing operating procedures (SOPs), the order, and or higher commander's guidance. The unit conducts the bypass without being detected, without being delayed by obstacles, or without becoming decisively engaged by the enemy force. The unit complies with the ROE, mission instructions, higher HQ order, and other special orders. All communication and reporting is according to applicable SOPs.

TASK STEPS AND PERFORMANCE MEASURES	GO	NO-GO

PLAN
1.   The unit leader receives an OPORD or a fragmentary order (FRAGO) that requires the unit to conduct a bypass. He issues a warning order (WARNO) in enough time for element leaders to have maximum planning time. The WARNO must include—
    a. Tentative unit organization for the bypass.
    b. Location and tentative timeline for the bypass, including the no later than time for beginning the movement.
    c. Guidance directing the unit to conduct rehearsals; any initial movement; initiate reconnaissance tasks and commander's critical information requirements (CCIRs).
*2. The unit leader develops a tentative plan according to troop-leading procedures (TLPs). He takes the following actions:		

TASK STEPS AND PERFORMANCE MEASURES	GO	NO-GO
a. Conducts mission analysis, using company intelligence support team (COIST), focusing on the mission given, enemy forces and their capabilities, terrain and weather effects, troops available, time available to execute the operation, and civil considerations (METT-TC); taking advantage of maps, imagery, unmanned aircraft systems (UASs), unattended ground sensors (UGSs), and other available capabilities.		
b. Develops a tentative course of action. Takes the following actions:		
(1) Identifies likely enemy avenues of approach.		
(2) Identifies bypass route.		
(3) Identifies tentative security and support by fire positions.		
(4) Marks tentative dismount points on maps, as appropriate.		
c. Identifies direct fire responsibilities and requests indirect fire, CAS, and attack aviation support according to the higher HQ's intent.		
d. Organizes the unit as needed to accomplish the mission and or compensate for combat losses. Takes the following actions:		
(1) Designates a fixing force to maintain contact with the enemy and assist the remainder of the unit during the bypass.		
(2) Designates other elements of the unit, as needed.		
e. Develops contingency plans for actions on contact with the enemy, casualty evacuation (CASEVAC), and crossing of danger areas as required.		
f. Conducts composite risk management to identify, assess, develop, and implement controls for hazards and to mitigate associated risks. (Refer to Task 07-2-5063, Conduct Composite Risk Management [Platoon-Company].)		
PREPARE		
3. The unit prepares for the bypass. It takes the following actions:		
a. Conducts rehearsals, if possible.		
b. Completes final inspections.		

TASK STEPS AND PERFORMANCE MEASURES	GO	NO-GO

*4. The leader supervises subordinate TLPs to ensure planning and preparations are on track and consistent with the unit leader's intent.
5. The unit issues FRAGOs as needed to address changes to the plan identified during TLPs and/or rehearsals.
EXECUTE
*6. The unit leader or designated representative conducts leader reconnaissance. He takes the following actions:
    a. Determines nature of enemy or obstacle contact, including size, location, composition, and other factors.
    b. Identifies bypass route that affords adequate cover and concealment and/or intervening distance, preventing the enemy from effectively obstructing and or engaging the unit.
    c. Plans additional security measures, such as employment of screening or obscuring smoke.
    d. Ensures bypass affording routes away from enemy positions, obstacles, and enemy engagement areas.
    e. Ensures that terrain along the bypass supports the maneuver of the unit and follow-on force.
    f. Adjusts the plan based on updated intelligence and reconnaissance effort.
    g. Disseminates updated reports, overlays, and other pertinent information.
*7. The unit leader issues the OPORD and uses FRAGOs as needed to redirect actions of subordinate elements.
8. The fixing element gains and maintains contact with the enemy. It—
    a. May not have to use direct fire.
    b. Reports enemy actions according to SOPs.
9. The unit conducts tactical movement or maneuver along the route or axis. It takes the following actions:
    a. Uses appropriate movement techniques and formations.
    b. Maintains proper weapons orientation to ensure 360-degree security.
    c. Identifies and reacts to enemy forces along the route. Takes the following actions:
        (1) Executes appropriate drills.
        (2) Completes the following as needed:
            (a) Armed vehicles suppresses on the move.		

TASK STEPS AND PERFORMANCE MEASURES	GO	NO-GO
(b) Mounted element members remains mounted.		
(c) Calls for and adjusts indirect fire and smoke to screen movement past the enemy position.		
(d) Reports the size and the location of the enemy to the higher HQ commander, and the unit continue the mission.		
d. Avoids detection and or delays, if possible.		
e. Avoids decisive engagement.		
f. Marks bypass according to the unit's SOPs.		
g. Once the rest of the unit clears the enemy position or obstacle, the fixing element—		
(1) Hand the enemy over to a support force breaks contact and rejoins the unit, according to the OPORD. OR		
(2) Remains attached to the follow-on forces, if applicable.		
*10. The unit leader directs maneuvers, as needed, to prevent becoming decisively engaged according to the OPORD when the element cannot bypass an enemy forces. He takes the following actions:		
a. Establishes a base of fire to suppress the enemy and prevent him from repositioning any part of his force. Takes the following actions:		
(1) Ensures vehicles seek covered positions and the mounted elements dismounts.		
(2) Suppresses the enemy using direct and indirect fires.		
b. Employs or calls for smoke to facilitate the maneuver of the rest of the unit.		
*11. The unit leader reports to higher HQ completion of the delay.		
ASSESS		
12.The unit consolidates and reorganization as needed. (Refer to Task 07-2-5027, Conduct Consolidation and Reorganization [Platoon–Company].)		
13.The unit continues operations as directed.		
* indicates a leader task step.		

SUPPORTING INDIVIDUAL TASKS

Task Number	Task Title
171-121-4045	Conduct Troop-Leading Procedures
171-610-0001	Perform a Map Reconnaissance
071-326-5630	Conduct Movement Techniques by a Platoon
171-121-3009	Control Techniques of Movement
301-348-1050	Report Information of Potential Intelligence Value

SUPPORTING COLLECTIVE TASKS

Task Number	Task Title
07-2-1342	Conduct Tactical Movement (Platoon-Company)
07-2-3000	Conduct Support by Fire (Platoon-Company)
07-2-3027	Integrate Direct Fires (Platoon-Company)
07-2-3036	Integrate Indirect Fire Support (Platoon-Company)
07-2-5009	Conduct a Rehearsal (Platoon-Company)
07-2-5027	Conduct Consolidation and Reorganization (Platoon-Company)
07-2-9006	Conduct a Passage of Lines as the Passing Unit (Platoon-Company)
07-3-9013	Conduct Actions on Contact
07-3-9017	Conduct Actions at Danger Areas
08-2-0003	Treat Casualties
08-2-0004	Evacuate Casualties
44-3-3220	Perform Passive Air Defense Measures
44-3-3221	Perform Active Air Defense Measures

SUPPORTING BATTLE/CREW DRILLS

Drill Number	Drill Title
17-3-D8008	React to an Obstacle
17-3-D9509	Break Contact
05-3-D0016	Conduct the 5 Cs

TASK: Conduct a Defense (Platoon-Company) (07-2-9003)

(FM 3-21.10)   (FM 3-21.8)   (FM 3-90.1)

CONDITIONS: The unit conducts operations as part of a higher headquarters (HQ) and receives an operation order (OPORD) or fragmentary order (FRAGO) to defend at the location and time specified. The defense may be conducted utilizing the techniques of sector, battle positions, strong point, or perimeter defense. Time is available for a deliberate occupation of defensive positions. All necessary personnel, equipment, indirect fire, and close air support (CAS) are available. The unit has communications with higher, adjacent, and subordinate elements. The unit receives guidance on the rules of engagement (ROE) and may also have specific mission instructions, such as a peace mandate, terms of reference, and a status-of-forces agreement (SOFA). Military and civilian, joint and multinational partners, and news media may be present in the operational environment (OE). Some iterations of this task should be performed in mission-oriented protective posture 4 (MOPP 4).

STANDARDS: The unit defends according to the standing operating procedures (SOPs), the order, and/or higher commander's guidance. The unit occupies designated defensive positions, covers designated portion of the engagement area (EA) or sector of fire, and maintains security. The unit destroys or defeats the enemy force within the assigned area. The unit complies with the ROE, mission instructions, higher HQ orders, and other special orders.

TASK STEPS AND PERFORMANCE MEASURES	GO	NO-GO
PLAN		
*1. Unit leaders receive an operation order (OPORD) or a fragmentary order (FRAGO) directing the unit to conduct a defense. They take the following actions:		
a. Conduct an initial assessment using the elements of mission, enemy, terrain and weather, troops and support available, time available, and civil considerations (METT-TC).		
b. Develop a planning and preparation timeline for the defense.		
2. Unit leaders issue a warning order (WARNO) to element leaders ensuring that subordinate leaders have sufficient time for their own planning and preparation needs. The WARNO must include—		
a. Type of defense.		
b Tentative unit organization for the defense.		

TASK STEPS AND PERFORMANCE MEASURES	GO	NO-GO

c.   Tentative location of defensive positions.
    d.   Tentative timeline for the operation, including tentative movement times and the no later than defend time.
    e.   Guidance on movement; initial surveillance and reconnaissance tasks and responsibility for the commander's critical information requirements (CCIRs).
*3. Unit leaders develop a tentative plan according to troop-leading procedures (TLPs). (Refer to Task 07-2-5081, Conduct Troop-Leading Procedures [Platoon-Company] for more information.) They take the following actions:
    a.   Conduct mission analysis by using the company intelligence support team (CoIST) focusing on METT-TC; taking advantage of maps, imagery, unmanned aircraft systems (UASs), unattended ground sensors (UGSs), and other available capabilities.
    b.   Develop a course of action. Take the following actions:
        (1) Identify the most likely enemy avenues of approach.
        (2) Identify the enemy scheme of maneuver using intelligence products.
        (3) Determine where to kill the enemy.
        (4) Plan for emplacement and integration of obstacles and direct and indirect fire weapon systems.
        (5) Plan for reconnaissance and rehearsal of actions in the EA.
    c.   Develop a defensive plan. Take the following actions:
        (1) Identify a tentative EA.
        (2) Develop an initial unit fire plan.
        (3) Identify existing and supporting manmade obstacles to force the enemy into the EA.
        (4) Identify tentative primary, alternate, and supplemental fighting positions.
        (5) Identify tentative observation post (OP) and sensor positions.
        (6) Integrate indirect fires, CAS, and attack aviation according to the higher HQ fire support plan.
        (7) Integrate smoke and obscuration.		

TASK STEPS AND PERFORMANCE MEASURES	GO	NO-GO
(8) Develop casualty evacuation plan.		
(9) Identify tentative dismount/remount points (as required).		
(10) Designate fire control measures.		
(11) Develop disengagement criteria.		
(12) Organize the unit as needed to accomplish the mission identifying the decisive, shaping, and supporting efforts for all phases of the defense.		
(13) Finalize positions and obstacle locations.		
(14) Complete the plan.		
d. Conduct composite risk management to identify, assess, develop, and implement controls for hazards and to mitigate associated risks. (Refer to Task 07-2-5063, Conduct Composite Risk Management [Platoon-Company] for more information.)		
*4. Unit leaders issue the OPORD and use FRAGOs as needed to redirect actions of subordinate elements.		
PREPARE		
5. The unit starts movement to a tactical assembly area or designated area short of the defensive positions.		
*6. Unit leaders and the reconnaissance element conduct the final reconnaissance (based on factors of METT-TC). They take the following actions:		
a. Pinpoint the defensive positions.		
b. Position security elements.		
c. Determine and confirm the EA.		
d. Drive or walk the EA to confirm the selected positions and establish target reference points (TRPs).		
e. Ensure positions are free of enemy, mines, and obstacles.		
f. Select primary, alternate, supplementary, and subsequent fighting positions (mounted and dismounted) to achieve the desired effect for each EA.		
g. Designate the hide positions for each battle position.		
h. Confirm location(s) of obstacles.		
i. Assign the elements AOs and OP locations. (OPs should have wire communications, if available.)		

TASK STEPS AND PERFORMANCE MEASURES	GO	NO-GO

j.   Designate the location for the command post (CP), early warning systems, and the chemical alarm systems (if assigned).
    k.   Identify dead space between elements and determined how best to cover the dead space.
    l.   Identify weapon systems positions so the required number of weapons, vehicles (as applicable), and elements can effectively cover each EA and avenues of approach.
    m. Set engagement priorities for each direct fire weapon system.
    n.   Plan for the fire control techniques to allow the unit to focus and redistribute fires into the EA.
    o.   Select covered and concealed routes between primary, alternate, and supplementary defensive positions.
    p.   Confirm tentative dismount/remount points (as applicable).
    q.   Verify and update intelligence information.
    r.   Leave a surveillance team to observe the defensive positions, if required.
    s.   Return to the unit location.
*7. Unit leaders adjust the plan based on updated intelligence and reconnaissance effort.
*8. Unit leaders update the enemy situation.
*9. Unit leaders disseminate updated reports, overlays, and other pertinent information.
*10. Unit leaders or designated representatives conduct initial defense coordination with adjacent unit, focusing on the following requirements:
    a.   Locations of OPs and patrols.
    b.   Communication information.
    c.   Unit positions including locations of mission command nodes.
    d.   Routes to be used during occupation and repositioning.
    e.   Overlapping fires (to ensure that direct fire responsibility is clearly defined and dead space is covered).
    f.   TRPs.
    g.   Indirect fire information.
    h.   Air defense considerations, if applicable.
    i. Sustainment considerations.		

TASK STEPS AND PERFORMANCE MEASURES	GO	NO-GO
*11. Unit leaders use FRAGOs as needed to redirect actions of subordinate elements.		
*12. The unit moves tactically to assigned defensive areas and prepares to occupy battle positions (BP). It takes the following actions:		
a. Uses covered and concealed routes.		
b. Enforces camouflage, noise, light, and litter discipline.		
c. Maintains security during movement.		
*13. The unit establishes the defense. It takes the following actions:		
a. Posts local security.		
b. Positions key weapons systems, vehicles, and other assets to effectively cover each EA.		
c. Conducts reconnaissance of the EA from the enemy's perspective (if possible).		
d. Assigns sectors of fire, engagement priorities, and other fire control measures.		
e. Ensures the unit is tied in with the unit on its right and left.		
f. Designates final protective fires (FPF) and final protective lines (FPL).		
g. Clears fields of fire.		
h. Prepares range cards/sector sketches.		
i. Constructs primary defensive positions according to unit SOP and/or as directed.		
j. Establishes communications.		
k. Emplaces claymore mines and protective obstacles as required. Takes the following actions:		
(1) Identifies dead space and requirements to refine the location of the obstacle group and fire control measures.		
(2) Ensures obstacles are covered by direct or indirect fire and under friendly observation.		
(3) Ensures obstacles are concealed from enemy observation as much as possible.		
(4) Ensures obstacles are employed in depth.		
(5) Ensures obstacles are tied in with existing obstacles, if possible.		
l. Maintains security (to include OPs, hasty perimeter, or security patrols).		

TASK STEPS AND PERFORMANCE MEASURES	GO	NO-GO
*14. Unit leaders consolidate sketch cards and finalize the unit fire plan.		
*15. The unit conducts the following rehearsals of the defense as time permits:		
a. Leadership rehearsal of the engagement using brief back format.		
b. Full force rehearsal of the engagement.		
c. Rehearsal of displacement to alternate and supplemental positions and the withdrawal plan.		
*16. As time permits, unit leaders direct elements to take additional steps to improve positions according to unit SOPs. They take the following actions:		
a. Add overhead cover.		
b. Emplace camouflage, alarms, and decoys.		
c. Establish alternate and supplemental positions according to unit SOP and/or as directed.		
d. Stockpile ammunition, food, and water.		
e. Establish detainee, wounded-in-action (WIA), and killed-in-action (KIA) collection points.		
f. Complete vehicle maintenance and prepared pre-fire checks.		
g. Establish a sleep and rest plan.		
17. Unit leaders or designated representatives conduct final coordination with adjacent units, focusing on the following requirements:		
a. Unit positions.		
b. Locations of OPs and patrols.		
c. Alternate, supplementary, and subsequent BPs.		
d. Sectors of fire and observation overlap.		
e. Obstacles (location and type).		
*18. Unit leaders adjust readiness condition (REDCON) status according to METT-TC factors, OPORD or FRAGO, and unit SOPs using Force XXI Battle Command Brigade and Below (FBCB2), field manuals (FMs), or other tactical means. They take the following actions:		
a. Assess readiness requirements based on tactical situation and METT-TC factors.		
b. Direct unit to assume appropriate REDCON level.		
*19. Unit leaders coordinate and/or synchronize actions of subordinate elements.		

TASK STEPS AND PERFORMANCE MEASURES	GO	NO-GO
*20. Unit leaders use FRAGOs as needed to redirect actions of subordinate elements.		
EXECUTE		
21. The unit executes the defense. It takes the following actions:		
a. Scans AOs.		
b. Engages enemy forces. Takes the following actions:		
(1) Uses indirect fires and CAS until enemy reaches direct fire trigger line.		
(2) Initiates direct fire engagements on command/or when the engagement criteria is met.		
(3) Destroys or forces enemy withdraw from EA.		
(4) Reports contact to higher commander and adjacent units.		
(5) Employs reserve and/or counterattack according to METT-TC.		
c. Displaces as required or when displacement criteria is met.		
ASSESS		
22. The unit consolidates and reorganizes as needed. (Refer to Task 07-2-5027, Conduct Consolidation and Reorganization [Platoon-Company] for more information.)		
23. The unit continues operations as directed.		
*indicates a leader task step.		

SUPPORTING INDIVIDUAL TASKS

Task Number	Task Title
171-121-4045	Conduct Troop-Leading Procedures
171-610-0001	Perform a Map Reconnaissance
071-326-5770	Prepare a Platoon Sector Sketch
171-121-3009	Control Techniques of Movement
061-283-6003	Adjust Indirect Fire
301-348-1050	Report Information of Potential Intelligence Value
071-326-5703	Construct Individual Fighting Positions
071-313-3454	Engage Targets with a Caliber .50 M2 Machine Gun
052-191-1362	Camouflage Equipment
061-284-3040	Engage Targets with Close Air Support

SUPPORTING COLLECTIVE TASKS

Task Number	Task Title
07-2-1342	Conduct Tactical Movement (Platoon-Company)
07-2-1387	Employ a Reserve Force (Platoon-Company)
07-2-1396	Employ Obstacles (Platoon-Company)
07-2-3027	Integrate Direct Fires (Platoon-Company)
07-2-3036	Integrate Indirect Fire Support (Platoon-Company)
07-2-5009	Conduct a Rehearsal (Platoon-Company)
07-2-5027	Conduct Consolidation and Reorganization (Platoon-Company)
07-2-6045	Employ Camouflage, Concealment, and Deception Techniques (Platoon-Company)
07-2-9007	Conduct a Passage of Lines as the Stationary Unit (Platoon-Company)
08-2-0003	Treat Casualties
08-2-0004	Evacuate Casualties
44-3-3220	Perform Passive Air Defense Measures
44-3-3221	Perform Active Air Defense Measures

SUPPORTING BATTLE/CREW DRILLS

Drill Number	Drill Title
07-3-D9501	React to Contact (Visual, IED, Direct Fire [includes RPG])
17-3-D8004	React to Air Attack

TASK: Conduct a Delay (Platoon-Company) (07-2-9004)

(FM 3-21.10)   (FM 3-21.8)

CONDITIONS: The unit conducts operations as part of a higher headquarters (HQ) and receives an operation order (OPORD) or fragmentary order (FRAGO) to delay the enemy for a specific time at the location and time specified. The enemy can attack by air, indirect fire, and ground (mounted or dismounted). All necessary personnel, equipment, indirect fire and close air support (CAS) are available. The unit communicates with higher, adjacent, and subordinate elements. The unit receives guidance on the rules of engagement (ROE) and may also receive specific mission instructions, such as a peace mandate, terms of reference, and a status-of-forces agreement (SOFA). Military and civilian, joint and multinational partners, and news media may be present in the operational environment (OE). Some iterations of this task should be performed in mission-oriented protective posture (MOPP 4).

STANDARDS: The unit conducts the delay according to the unit standing operating procedures (SOPs), the order, and/or higher commander's guidance. The unit occupies initial delay positions, forces the enemy to slow their advance, complies with all control measures, and disengages from the enemy as directed. The unit does not engage decisively. The unit complies with the ROE, mission instructions, higher HQ orders, and other special orders. All communication and reporting is according to applicable SOPs.

TASK STEPS AND PERFORMANCE MEASURES	GO	NO-GO
PLAN		
*1. Unit leaders receive an OPORD or a fragmentary order FRAGO directing the unit to conduct a delay. The unit commander issues a warning order (WARNO) to element leaders ensuring that subordinate leaders have sufficient time for their own planning and preparation needs. The WARNO must include—		
a. Type of delay to conduct (in sector or forward of a line or position for a specified time).		
b. Tentative unit organization for the delay including—		
(1) Main body.		
(2) Security force.		
(3) Reserve.		

TASK STEPS AND PERFORMANCE MEASURES	GO	NO-GO
c. Tentative timeline for the operation including tentative movement times.		
d. Guidance directing the unit to conduct rehearsals and any initial movement, and initiate reconnaissance tasks and commander's critical information requirements (CCIRs).		
*2. Unit leaders begin developing a tentative plan according to the troop-leading procedures (TLP). (Refer to Task 07-2-5081, Conduct Troop-leading Procedures [Platoon-Company] for more information.)They take the following actions:		
a. Conduct mission analysis company intelligence support team (CoIST) focusing on the mission given, enemy forces and their capabilities, terrain and weather effects, troops available, time available to execute the operation, and civil considerations (METT-TC); taking advantage of maps, imagery, unmanned aircraft systems (UASs), unattended ground sensors (UGSs), and other available capabilities.		
b. Develop a tentative plan. Take the following actions:		
(1) Identify initial and subsequent delay positions.		
(2) Identify general routes between delay positions.		
(3) Identify tentative, security, support by fire, and assault positions, if required.		
(4) Identify likely enemy avenues of approach.		
(5) Plan obstacles to slow the enemy advance.		
(6) Establish the disengagement criteria.		
(7) Identify evacuation routes.		
(8) Identify key terrain.		
(9) Mark tentative dismount points on maps as appropriate.		
(10) Plan and coordinate indirect fire support and/or close air support, if available and incorporates the higher HQ fire plan.		

TASK STEPS AND PERFORMANCE MEASURES	GO	NO-GO
(11) Develop control measures to include unit-level phase lines and graphics.		
(12) Develop contingency plans for possible offensive operations.		
(13) Organize the unit as needed to accomplish the mission and/or compensate for combat losses.		
(14) Conduct composite risk management to identify, assess, develop, and implement controls for hazards and to mitigate associated risks. (Refer to Task 07-2-5063, Conduct Composite Risk Management [Platoon-Company] for more information.)		
*3. Unit leaders issue the OPORD and use FRAGOs as needed to redirect actions of subordinate elements.		
PREPARE		
4. The unit prepares for the delay. It takes the following actions:		
a. Conducts a reconnaissance and marks general routes between delay positions including—		
(1) Selects security, support, and assault positions, if required.		
(2) Leaves security at initial delay positions.		
(3) Verifies and update intelligence information.		
b. Constructs a series of alternate fighting positions.		
c. Stockpiles supplies, fuel, and ammunition in designated positions as required.		
d. Constructs obstacles as needed.		
e. Evacuates all unneeded personnel, supplies, and equipment.		
f. Prepares to destroy supplies and equipment that cannot be evacuated.		
g. Conducts rehearsal during daylight and periods of reduced visibility, if possible.		
h. Supervises subordinate troop-leading procedures to ensure planning and preparations are on track and consistent with the unit commander's intent.		

TASK STEPS AND PERFORMANCE MEASURES	GO	NO-GO
EXECUTE		
5. The unit executes the delay. It takes the following actions:		
a. Occupies initial delay positions according to the OPORD and/or guidance from higher HQ.		
b. Forces the enemy to slow their advance by forcing them to change their movement formations and speed by employing—		
(1) Ambushes.		
(2) Snipers.		
(3) Obstacles and minefields.		
(4) Direct and indirect fires.		
c. Complies with all control measures and time constraints specified in the OPORD.		
6. The unit does not become decisively engaged. It takes the following actions:		
a. Disengages from the enemy and withdraws to new positions before enemy assault.		
b. Continues delaying action until one of the following is met:		
(1) The delaying force conducts a rearward passage of lines through a defending force.		
(2) The delaying force reaches defensible terrain and transitions to the defense.		
(3) The advancing enemy force reaches a culminating point and can no longer continue to advance.		
(4) The delaying force goes on the offensive.		
*7. Unit leaders report to higher HQ the completion of the delay.		
ASSESS		
8. The unit consolidates and reorganizes as needed. (Refer to Task 07-2-5027, Conduct Consolidation and Reorganization [Platoon–Company] for more information.)		

TASK STEPS AND PERFORMANCE MEASURES	GO	NO-GO
9. The unit continues operations as directed.		
*indicates a leader task step.		

SUPPORTING INDIVIDUAL TASKS

Task Number	Task Title
171-121-4045	Conduct Troop-Leading Procedures
171-610-0001	Perform a Map Reconnaissance
071-326-5630	Conduct Movement Techniques by a Platoon
171-121-3009	Control Techniques of Movement
061-283-6003	Adjust Indirect Fire
301-348-1050	Report Information of Potential Intelligence Value
071-326-5606	Select an Overwatch Position
071-025-0007	Engage targets with an M249 Machine gun
071-450-0041	Conduct a Point Ambush
071-450-0036	Conduct an Antiarmor Area Ambush by a Platoon
071-313-3454	Engage Targets with a Caliber .50 M2 Machine

SUPPORTING COLLECTIVE TASKS

Task Number	Task Title
07-2-1252	Conduct an Antiarmor Ambush (Platoon-Company)
07-2-1342	Conduct Tactical Movement (Platoon-Company)
07-2-1396	Employ Obstacles (Platoon-Company)
07-2-3018	Employ Snipers (Platoon-Company)
07-2-3036	Integrate Indirect Fire Support (Platoon-Company)
07-2-5027	Conduct Consolidation and Reorganization (Platoon-Company)
07-2-9006	Conduct a Passage of Lines as the Passing Unit (Platoon-Company)
07-2-9009	Conduct a Withdrawal (Platoon-Company)
07-2-9010	Conduct an Ambush (Platoon-Company)
07-3-1072	Conduct a Disengagement
07-3-9013	Conduct Action on Contact
07-3-9017	Conduct Actions at Danger Areas
08-2-0003	Treat Casualties
08-2-0004	Evacuate Casualties
43-2-4522	Destroy Supplies and Equipment

SUPPORTING BATTLE/CREW DRILLS

Drill Number	Drill Title
07-3-D9501	React to Contact (Visual, IED, Direct Fire [includes RPG])
07-3-D9504	React to Indirect Fire

TASK: Conduct a Passage of Lines as the Passing Unit (Platoon-Company) (07-2-9006)

(FM 3-21.10)   (FM 3-21.8)

CONDITIONS: The unit conducts operations as part of a larger force and receives an operation order (OPORD) or fragmentary order (FRAGO) to conduct a forward or rearward passage of lines. The stationary unit has been identified. All necessary unit personnel and equipment are available. Indirect fire and close air support (CAS) are available. The unit has established communications with required units at all echelons. The unit has guidance on the rules of engagement (ROE). Coalition forces and noncombatants may be present in the operational environment. Some iterations of this task should be performed under the conditions of: mission, enemy, terrain and weather, troops and support available, time available, and civil considerations (METT-TC) that aid or limit performance, or in mission-oriented protective posture 4 (MOPP4).

STANDARDS: The unit conducts the passage of lines according to the standing operating procedures (SOPs), the order, and or higher commander's guidance. The unit completes necessary coordination with higher, adjacent, and stationary elements. The unit passes through the stationary unit with no compromise of security, and complies with the ROE, mission instructions, higher headquarters (HQ) order, and other special orders.

TASK STEPS AND PERFORMANCE MEASURES	GO	NO-GO

PLAN
*1. Unit leaders receive an OPORD or a FRAGO that requires the unit to conduct a passage of friendly lines and issue a warning order (WARNO) according to troop-leading procedures (TLPs) and unit SOPs. The WARNO must include—
    a. Location and tentative timeline for the passage of lines, including movement times and no later than time.
    b. Tentative unit organization during the passage of lines, identifying security and quartering parties as necessary.
    c. Guidance to conduct rehearsals and any initial movement, initiate surveillance and reconnaissance tasks, and ensure commander's critical information requirements (CCIRs).		

TASK STEPS AND PERFORMANCE MEASURES	GO	NO-GO

*2. Unit leaders develop a tentative plan according to the TLPs. They take the following actions:
 	
    a. Conduct mission analysis by using company intelligence support team (CoIST); focusing on METT-TC; taking advantage of maps, imagery; human intelligence (HUMINT); signal intelligence (SIGINT); unmanned aircraft systems (UASs); unattended ground sensors (UGSs); and other available capabilities.
 	
    b. Develop a tentative course of action. Take the following actions:
 	
        (1) Identify passage points and passage lanes (primary and alternate).
 	
        (2) Identify likely enemy avenues of approach.
 	
        (3) Identify security and support positions, if applicable.
 	
        (4) Mark tentative dismount points on maps as appropriate.
 	
        (5) Plan and coordinate indirect fire support and or close air support, if available.
 	
        (6) Identify and assign direct fire responsibilities.
 	
        (7) Organize the unit as necessary to accomplish the mission and or compensate for combat losses.
 	
    c. Develop contingency plans on chance contact with the enemy before, during, and after the passage, actions on break down of vehicles during passage, and casualty evacuation (CASEVAC) during the operation.
 	
    d. Conduct risk management to identify, assess, develop, and implement controls for hazards and mitigate associated risks.
 	
*3. Unit leaders or designated representatives coordinate with the stationary unit and exchange and/or coordinate the following information:
 	
    a. Updated enemy situation.
 	
    b. Friendly situation and disposition.
 	
    c. Signal operating instructions information.
 	
    d. Command post location.
 	
    e. Contact points (primary, alternate).
 	
    f. Number and type of personnel/vehicles involved in the passage.		

TASK STEPS AND PERFORMANCE MEASURES	GO	NO-GO
g. Estimated time of arrival of passing elements.		
h. Recognition signals.		
i. Verification and/or designation of fire coordination measures.		
j. Verification of the command relationship with the stationary unit.		
k. Verification of known obstacle types and locations, and applicable breach locations, passage points, or bypass routes (friendly, existing, and enemy).		
l. Supporting fires information, to include available assets, smoke data, and target numbers and locations.		
m. Passage lanes data, to include alternate routes, start point, release point, passage points, and checkpoints.		
n. Location and number of guides and number and type of vehicles.		
o. Time of passage.		
p. Rally points (both near and far) and assembly areas.		
q. Line of departure.		
r. Battle handover line (BHL) and battle handover criteria for the transfer of responsibility for the control of the sector takes place, if applicable.		
s. Sustainment information, including the following:		
(1) Resupply of Classes III and V.		
(2) Medical evacuation assets.		
(3) Handling of enemy prisoners of war.		
(4) Maintenance requirements and available assets.		
t. Action on contact if enemy is encountered during the passage.		
u. Verification of actions to take place following coordination of the passage.		
*4. Unit leaders or designated representatives complete coordination and preparations for the passage. They take the following actions:		
a. Conduct tactical movement back to the unit position, if applicable.		

TASK STEPS AND PERFORMANCE MEASURES	GO	NO-GO
b. Pass all pertinent information and/or FRAGO to the unit.		
c. Report all pertinent information from the coordination to the higher commander if acting as liaison for higher HQ.		
d. Direct subordinate element leaders to complete troop-leading procedures required to plan the passage.		
PREPARE		
*5. Unit leaders issue the OPORD and use FRAGOs as necessary to redirect actions of subordinate elements.		
6. The unit conducts a rehearsal.		
*7. Unit leaders coordinate/synchronize actions of subordinate elements.		
*8. Unit leaders use FRAGOs as needed to redirect actions of subordinate elements.		
EXECUTE		
9. The passing unit moves to an assembly area or an attack position.		
10. Designated liaison personnel link up with guides and confirm coordination information with stationary unit.		
11. The unit conducts tactical movement to the passage point. It takes the following actions:		
a. Establishes communications with stationary unit. Takes the following actions:		
b. Uses covered and concealed routes to the maximum extent possible.		
c. Uses best formation and movement technique based on factors of METT-TC.		
d. Maintains proper weapons orientation to ensure 360-degree security based on the formation selected.		
e. Can employ additional fire control measures to minimize the risk of fratricide.		
12. The unit moves through the passage point. It takes the following actions:		
a. Keeps communications with stationary unit to a minimum.		
b. Displays designated recognition signal on all vehicles, if applicable.		
c. Reports arrival time at the passage point to higher HQ.		

TASK STEPS AND PERFORMANCE MEASURES	GO	NO-GO

d. Passes through the passage point without halting or blocking it.
13. The unit moves along the passage lane. It takes the following actions:
   a. Conducts tactical movement through the passage lane.
   b. Orients weapon systems in the direction of known or suspected enemy contact.
   c. Follows directions given by guides at traffic control points.
   d. Uses alternate lanes if situation dictates.
*14. Unit leaders or representatives keep higher HQ informed. They take the following actions:
   a. Report graphic control measures for passage.
   b. Send situation report as needed during execution.
   c. Report completion of the passage.
ASSESS
15. If the passage is forward, the unit crosses the BHL and continues the mission.
16. If the passage is rearward, the unit crosses the BHL and takes the specified following actions:
   a. Moves to the location designated in the OPORD without halting or blocking the passage lane.
   b. Occupies an assembly area or continues on assigned mission as specified in the OPORD.
*indicates a leader task step.		

SUPPORTING INDIVIDUAL TASKS

Task Number	Task Title
07-2-5036	Conduct Coordination (Platoon-Company)
113-637-2001	Communicate Via a Tactical Radio in a Secure Net
052-192-3262	Prepare for an Improvised Explosive Device (IED) Threat Prior to Movement (Unclassified/For Official Use Only) (U//FOUO)
052-703-9107	Plan for an Improvised Explosive Device (IED) Threat in a COIN Environment (Unclassified/For Official Use Only) (U//FOUO)
052-192-1271	Identify Visual Indicators of an Improvised Explosive Device (IED) (Unclassified/For Official Use Only) (U//FOUO)

071-326-5505	Issue an Operation Order at the Company, Platoon, or Squad Level
071-326-5502	Issue a Fragmentary Order
071-329-1030	Navigate from One Point on the Ground to Another Point While Mounted
071-329-1006	Navigate from One Point on the Ground to Another Point While Dismounted

SUPPORTING COLLECTIVE TASKS

Task Number	Task Title
07-2-1342	Conduct Tactical Movement (Platoon-Company)
07-2-5036	Conduct Coordination (Platoon-Company)
07-2-5081	Conduct Troop-Leading Procedures (Platoon-Company)
07-2-9005	Conduct a Linkup (Platoon-Company)
07-3-9013	Conduct Action on Contact

SUPPORTING BATTLE/CREW DRILLS

Drill Number	Drill Title
07-3-D9501	React to Contact (Visual IED, Direct Fire [includes RPG])
17-3-D8004	React to Air Attack

TASK: Conduct a Passage of Lines as the Stationary Unit (Platoon - Company) (07-2-9007)

(FM 3-21.10)   (FM 3-21.8)

CONDITIONS: The unit conducts operations as part of a larger force and receives an operation order (OPORD) or fragmentary order (FRAGO) to pass another element through their lines. All necessary unit personnel and equipment are available. Indirect fire and close air support (CAS) are available. The unit has established communications with required units at all echelons and has guidance on the rules of engagement (ROE). Coalition forces and noncombatants may be present in the operational environment. Civilians, government agencies, nongovernment organizations, and local and international media may be in the area. Some iterations of this task should be performed under mission, enemy, terrain and weather, troops available, time available, and civil considerations (METT-TC) conditions that aid or limit performance. Some iterations of this task should be performed in mission-oriented protective posture 4 (MOPP 4).

STANDARDS: The unit passes another element through their lines according to unit standing operating procedures (SOP), the order, and/or higher commander's guidance. The unit performs the necessary coordination/liaison with the passing element, designates and briefs guide personnel, passes another element through their lines, and conducts a battle/reconnaissance handover line (BHL/RHL). Unit complies with the ROE, mission instructions, higher headquarters order, and other special orders. Unit treats local inhabitants with respect.

TASK STEPS AND PERFORMANCE MEASURES	GO	NO-GO
PLAN		
*1. Unit leaders gain and or maintain situational understanding using available communications equipment, maps, intelligence summaries; situation reports (SITREPs) and other available information sources. Intelligence sources include company intelligence support team (CoIST), human intelligence (HUMINT), signal intelligence (SIGINT), and imagery intelligence (IMINT) to include unmanned aircraft systems (UAS) and unattended ground sensors (UGSs).		
2. Unit leaders receive an OPORD or FRAGO that requires their unit to pass a friendly unit through its lines and issue a warning order (WARNO) to the unit according to unit SOPs. The WARNO must include—		

TASK STEPS AND PERFORMANCE MEASURES	GO	NO-GO
a. Tentative location and timeline for the passage of lines, including movement times and no later than time.		
b. Tentative unit organization during the passage of lines, over watch elements, and guides.		
c. Guidance directing the unit to conduct rehearsals and initiate any movement.		
*3. Unit leaders develop a tentative plan according to the troop-leading procedures. They take the following actions:		
a. Conduct mission analysis focusing on the METT-TC; taking advantage of maps, imagery, HUMINT, SIGINT, unmanned aircraft UAS, UGS, and other available capabilities.		
b. Develop a tentative course of action. Take the following actions:		
(1) Identify passage point, if not specified by higher headquarters (HQ).		
(2) Identify contact points.		
(3) Identify an assembly area or attack position for staging of passing unit.		
(4) Develop direct and indirect fire responsibilities and control measures during conduct of the passage.		
c. Conduct risk management to identify, assess, develop, and implement controls for hazards and to mitigate associated risks.		
d. Organize as necessary to accomplish the mission and or compensate for combat losses.		
e. Coordinate with higher HQ and/or the passing unit. Take the following actions:		
(1) Obtain the following:		
(a) Location of passage point, if specified by higher HQ.		
(b) Signal operating instructions (SOI) information.		
(c) Passing unit designation.		
(d) Number/type of vehicles involved in the passage.		
(e) Estimated time of arrival of unit and time of passage.		

TASK STEPS AND PERFORMANCE MEASURES	GO	NO-GO

(f)   Order of march.
       (g)   Recognition signals.
       (h)   Guide requirements.
    (2) Provide the following as appropriate:
       (a)   Friendly situation and or disposition.
       (b)   Updated enemy situation.
       (c)   Terrain analysis.
       (d)   Supporting fires information to include available assets, smoke data, and target numbers and or locations.
       (e)   Locations of friendly obstacles and applicable breaching information, to include any routes through obstacles.
       (f)   Rally points and assembly area locations.
       (g)   Location of the line of departure.
       (h)   Location of the BHL/RHL and time the transfer of responsibility for the control of the sector will take place.
       (i)   Friendly unit locations.
       (j)   Information for resupply of Classes III and V.
       (k)   Information for medical evacuation assets.
       (l)   Information for handling of enemy prisoners of war.
       (m) Information for maintenance requirements and available assets.
PREPARE
*4. Unit leaders issue the OPORD and use FRAGOs as necessary to redirect actions of subordinate elements.
5.   Designated unit leadership briefs guide personnel on duties.
6.   The unit conducts rehearsal.
*7. Unit leaders issue FRAGOs, as needed, to address changes to the plan identified during the rehearsal.
EXECUTE
8.   Guide personnel move to appropriate positions and begin activities to support the passage. They take the following actions:
    a. Establish overwatch positions.		

TASK STEPS AND PERFORMANCE MEASURES	GO	NO-GO

b.   Reconnoiter and mark route for the passing unit.
 9.   Guide personnel reconnoiter and open passage lane. They take the following actions:
     a. Ensure lane provides adequate maneuver space for all passing unit vehicles/personnel.
     b. Assume positions to provide all-round defense for the passage.
 10. Guide personnel perform linkup with the passing unit and brief unit leaders as needed on the following:
     a.   Executes near and far recognition signals.
     b.   Verifies restrictive fire line and BHL as necessary.
     c.   Overwatches the passage and provides security as needed.
  11. The unit conducts physical linkup with passing element at the designated passage point and guides the passing unit through the passage lane to the release point.
  12. The unit conducts physical link-up with passing element at the designated passage point and guides the passing unit through the passage lane to the release point. (07-2-9005)
 ASSESS
 *13.The unit closes passage lane and any lanes through obstacles as required.
  14. Unit leaders report the completion of the passage to higher HQ.
 *indicates a leader task step.		

SUPPORTING INDIVIDUAL TASKS

Task Number	Task Title
071-410-0010	Conduct a Leader's Reconnaissance
071-326-5502	Issue a Fragmentary Order
071-331-0801	Challenge Persons Entering Your Area
071-326-5505	Issue an Operation Order at the Company, Platoon, or Squad Level

SUPPORTING COLLECTIVE TASKS

Task Number	Task Title
07-2-1324	Conduct Area Security (Platoon-Company)
07-2-3000	Conduct Support by Fire (Platoon-Company)
07-2-5036	Conduct Coordination (Platoon-Company)

07-2-6063	Maintain Operations Security (Platoon-Company)
07-2-9005	Conduct a Linkup (Platoon-Company)
07-3-9013	Conduct Action on Contact
07-2-5081	Conduct Troop-leading Procedures (Platoon-Company)

SUPPORTING BATTLE/CREW DRILLS

Drill Number	**Drill Title**
07-3-D9501	React to Contact (Visual IED, Direct Fire [includes RPG])

TASK: Conduct a Withdrawal (Platoon-Company) (07-2-9009)

(FM 3-21.10)   (FM 3-21.8)

CONDITIONS: The unit conducts operations as part of a higher headquarters (HQ) and receives an operation order (OPORD) or fragmentary order (FRAGO) directing it to disengage and withdraw immediately from the enemy and reposition for another mission. The withdrawal may or may not be conducted under enemy pressure and is unassisted. Indirect fire and close air support (CAS) are available. The unit may be directed to designate a detachment left in contact. All necessary unit personnel and equipment are available. The unit has established communications with required units at all echelons. The unit has received guidance on the rules of engagement (ROE). Coalition forces and noncombatants may be present in the operational environment. Some iterations of this task should be performed in mission-oriented protective posture 4 (MOPP 4).

STANDARDS: The unit conducts the withdrawal according to standing operating procedures (SOPs), the order, and or the higher commander's guidance. The unit disengages and moves to a designated location where the enemy cannot observe or engage it with direct fire. The unit designates a detachment left in contact. The unit complies with the ROE, mission instructions, higher HQ orders, and other special orders.

TASK STEPS AND PERFORMANCE MEASURES	GO	NO-GO

PLAN
*1. The unit leader receives an OPORD or a FRAGO that requires the unit to conduct a withdrawal. It issues a warning order (WARNO) to the unit according to troop-leading procedures (TLPs) and unit SOPs. The WARNO must include—
    a. Tentative timeline for the operation, including the latest time for beginning the withdrawal.
    b. Tentative unit organization for the operation.
    c. Guidance directing the unit to conduct rehearsals; any initial movement; initiate surveillance, reconnaissance tasks, and commander's critical information requirements (CCIRs).
*2. The unit leader begins developing a tentative plan according to TLPs. He takes the following actions:		

TASK STEPS AND PERFORMANCE MEASURES	GO	NO-GO
a. Gains and/or maintains situational understanding using available communications equipment, maps, intelligence summaries, situation reports (SITREPs), and other available information sources. Intelligence sources include company intelligence support team (CoIST), human intelligence (HUMINT), signal intelligence (SIGINT), and imagery intelligence (IMINT) to include unmanned aircraft systems (UASs) and unattended ground sensors (UGSs).		
b. Develops a tentative course of action. Takes the following actions:		
(1) Indicates the method of disengagement based on enemy pressure (not under pressure or under enemy pressure).		
(2) Determines when and where the withdrawal will start.		
(3) Identifies possible key terrain and routes based on the higher unit's graphics and his map.		
(4) Determines the locations for assembly areas or battle positions (BPs) to which the unit will withdraw.		
(5) Plans deception activities.		
(6) Conducts fire planning for direct fires, indirect fires and CAS along the withdrawal route to aid in the withdrawal.		
(7) Develops criteria for special instructions on employment of special weapons element (for example, mortars if available).		
(8) Designates the withdrawal task organization into security force, main body, and reserve, as needed, to accomplish the mission and or compensate for combat losses. Takes the following actions:		
(a) Designates an element of the security force as the detachment left in contact (DLIC).		
(b) Determines the size, composition, mission, and leader of the DLIC.		
(c) Designates a quartering party.		
c. Determines the withdrawal and linkup plan for the detachment left in contact.		
d. Plans to withdraw under limited visibility conditions, if possible.		

TASK STEPS AND PERFORMANCE MEASURES	GO	NO-GO
e. Develops contingency plans for chance contact, medical evacuation (MEDEVAC), and other events identified during mission analysis.		
f. Coordinates with adjacent units as necessary.		
g. Conducts risk management to identify, assess, develop, and implement controls for hazards and to mitigate associated risks.		
3. The quartering party takes the following actions:		
a. Conducts reconnaissance of positions to which unit will withdraw.		
b. Selects subordinate element positions and/or sectors.		
c. Selects observation post (OP) positions for unit.		
d. Provides guides as needed.		
e. Coordinates with the unit/element through which the unit will conduct a rearward passage of lines, if required.		
PREPARE		
*4. The unit leader issues the OPORD and uses FRAGOs as needed to redirect actions of subordinate elements.		
5. The unit conducts a rehearsal if withdrawal is not under enemy pressure. It walks the routes during daylight and limited visibility based on the factors of mission, enemy, terrain and weather, troops and support available, time available, and civil considerations (METT-TC).		
EXECUTE		
6. The unit withdraws under enemy pressure. It takes the following actions:		
a. Uses DLIC to cover the withdrawal of the main body by deception and maneuver. The unit leader uses one of the following methods to designate the DLIC:		
(1) Designates one element to execute the DLIC mission.		
(2) Constitutes DLIC using ad hoc elements of the main body with a designated leader.		
c. Moves main body from their positions to their designated assembly area/BP while covered by the DLIC, and then to the unit AA upon order or at the designated time (after all equipment and personnel are accounted for).		

TASK STEPS AND PERFORMANCE MEASURES	GO	NO-GO
d. Moves unit to the higher HQ assembly area upon order or at the designated time (after all equipment and personnel were accounted for). 7. The unit withdraws not under enemy pressure. It takes the following actions: a. Moves unneeded vehicles and/or equipment to the rear before the withdrawal started. b. Moves special weapon systems where they can provide support to cover the withdrawal. c. Withdraws least heavily engaged element first. Takes the following actions: (1) Ensures the DLIC disengages and moves into a position where it can best overwatch the disengagement of the more heavily engaged elements. (2) Ensures the main body continues maneuvering to the rear and provides overwatch in turn. 8. The unit completes the withdrawal. ASSESS 9. The unit consolidates and reorganizes as needed. 10. The unit continues operations as directed. * indicates a leader task step.		

SUPPORTING INDIVIDUAL TASKS

Task Number	Task Title
071-326-5606	Select an Overwatch Position
071-410-0016	Conduct Occupation of an Overwatch Position
071-410-0010	Conduct a Leader's Reconnaissance
071-410-0020	Plan for Use of Supporting Fires

SUPPORTING COLLECTIVE TASKS

Task Number	Task Title
07-2-1324	Conduct Tactical Movement (Platoon-Company)
07-2-1387	Employ a Reserve Force (Platoon-Company)
07-2-3018	Employ Snipers (Platoon-Company)
07-2-3036	Integrate Indirect Fire Support (Platoon-Company)
07-2-9006	Conduct a Passage of Lines as the Passing Unit (Platoon-Company)
07-2-9014	Occupy an Assembly Area (Platoon-Company)
07-3-9013	Conduct Action on Contact
07-3-9016	Establish an Observation Post
07-3-9017	Conduct Actions at Danger Areas

08-2-0003	Treat Casualties
08-2-0004	Evacuate Casualties
44-3-3221	Perform Active Air Defense Measures

SUPPORTING BATTLE/CREW DRILLS

Drill Number	Drill Title
07-3-D9501	React to Contact (Visual, IED, Direct Fire [includes RPG])
07-3-D9504	React to Indirect Fires
07-3-D9505	Break Contact

TASK: Conduct a Relief in Place (Platoon-Company) (07-2-9012)

(FM 3-21.10)   (FM 3-21.8)

CONDITIONS: The unit conducts operations as part of a higher headquarters (HQ) and receives an operation order (OPORD) or fragmentary order (FRAGO) to conduct a relief in place at the location and time specified. All necessary unit personnel and equipment are available. Indirect fire and close air support (CAS) are available. The unit has established communications with required units at all echelons. The unit has received guidance on the rules of engagement (ROE). Coalition forces and noncombatants may be present in the operational environment. Some iterations of this task should be performed in mission-oriented protective posture 4 (MOPP4).

STANDARDS: The unit conducts the relief in place according to the unit standing operating procedures (SOPs), the order, and/or higher commander's guidance. The unit conducts the necessary coordination, moves tactically to the designated contact point, occupies relieved unit's positions, and assumes responsibility for the fight without allowing the enemy an advantage. The unit complies with the ROE, mission instructions, higher HQ order, and other special orders.

TASK STEPS AND PERFORMANCE MEASURES	GO	NO-GO
PLAN		
*1. Unit leaders receive an OPORD or a FRAGO that requires the unit to conduct a relief in place (RIP), and issue a warning order (WARNO) to the unit according to troop-leading procedures (TLPs) and unit SOPs. The WARNO must include—		
a. Location and tentative timeline for the operation, including the no later than time for completion of the RIP.		
b. Tentative unit organization for the RIP.		
c. Guidance directing the unit to conduct rehearsals; any initial movement; initiate surveillance, and reconnaissance tasks and commander's critical information requirements (CCIRs).		
*2. Unit leaders develop a tentative plan according to TLPs. They take the following actions:		

TASK STEPS AND PERFORMANCE MEASURES	GO	NO-GO
a. Conduct mission analysis company intelligence support team (CoIST), focusing on the mission given, enemy forces and their capabilities, terrain and weather effects, troops available, time available to execute the operation, and civil considerations (METT-TC); taking advantage of maps, imagery, unmanned aircraft systems (UASs), unattended ground sensors (UGSs), and other available capabilities.		
b. Unit leaders make contact with the unit leaders of the counterpart unit, and develop a tentative course of action, which includes—		
(1) Identify contact points if not identified in the higher OPORD.		
(2) Identify routes to and from contact points.		
(3) Mark tentative dismount points on maps, as appropriate.		
(4) Identify direct fire responsibilities and integrating indirect fire support and or close air support, if necessary.		
(5) Plan for increased supply consumption during RIP.		
(6) Develop contingencies for chance contact with the enemy; identify command and control during the RIP.		
(7) Conduct risk management to identify, assess, develop, and implement controls for hazards to mitigate associated risks, when possible.		
*3. Unit leaders, or designated representatives, coordinate and/or exchange information with the relieved unit according to SOPs, the OPORD or FRAGO, and/or commander's guidance by coordinating and/or exchanging the following:		
a. Update enemy situation.		
b. Outgoing unit's tactical plan, to include: graphics, fire plans, and individual vehicles' and/or sector sketches.		
c. Location of vehicle and/or individual fighting positions (to include: hide, alternate, and supplementary positions).		
d. Fire support coordination, including: indirect fire plans and the time of relief for supporting artillery and mortar units.		

TASK STEPS AND PERFORMANCE MEASURES	GO	NO-GO
e. Types of weapon systems being replaced.		
f. Location and disposition of obstacles and the time responsibility will be transferred.		
g. Counterattack plans.		
h. Plans for other tasks the elements may have been tasked to perform.		
i. Supplies and equipment to be transferred.		
j. Movement control, route priority, and placement of guides.		
k. Sustainment support and evacuation, if necessary, for disabled vehicles.		
l. Time, sequence, and method of relief (relief of one unit at a time, simultaneous relief of units, and relief by occupation in depth or occupation of adjacent positions.		
m. Communications data information, which includes—		
(1) Frequencies.		
(2) Filters for digital equipment.		
(3) Signals.		
(4) Challenge and password.		
n. Battle hand over line (BHL) procedures for artillery.		
*4. Unit leaders or designated representatives and reconnaissance elements, conduct the reconnaissance based on METT-TC, which includes the following actions:		
a. Reconnoiter routes into and out of the position.		
b. Reconnoiter any assembly areas to be used.		
c. Reconnoiter logistics release points (LRPs).		
d. Reconnoiter primary, alternate, and supplementary positions.		
e. Reconnoiter obstacles.		
f. Reconnoiter patrol routes (primary and alternate) and observation post (OPs) locations, when possible.		
g. Verify and update priority intelligence requirements.		
h. Post security, if required.		
PREPARE		
*5. Unit leaders issue the OPORD and use FRAGOs, as needed, to redirect actions of subordinate elements.		

TASK STEPS AND PERFORMANCE MEASURES	GO	NO-GO
6. The unit conducts rehearsal, at a minimum, executing a leader's rehearsal in back brief format. EXECUTE		
7. The unit executes the RIP, while taking the following actions:		
a. Maintains operations security (OPSEC) by—		
(1) Changing frequencies on all element radios to the frequencies of the relieved unit.		
(2) Maintaining radio listening silence (if specified in the OPORD and or FRAGO).		
(3) Adhering to noise, light, and litter discipline.		
b. Initiates movement by—		
(1) Moving to predetermined contact points and meets guides from the relieved unit.		
(2) Establishing a command post (CP) with relieved unit CP.		
(3) Moving into hide positions and coordinating any final information.		
c. Occupies, as needed, preliminary positions in preparation for conducting the relief. The following procedures apply:		
(1) Occupy positions behind the unit to be relieved, as determined in the coordination or as specified in the OPORD.		
(2) Report to the higher HQ commander when occupation of preliminary position is complete and the element is prepared to conduct the relief.		
d. Completes preparations and coordination with the relieving or relieved unit. Takes the following actions:		
(1) Updates the enemy situation.		
(2) Completes transfer of sector sketches and fire plans.		
(3) Completes transfer of obstacle target folders and hasty protective minefield forms.		
e. Conducts the relief and transfer equipment and supplies, as required. This includes—		
(1) Conducts battle hand over with relieving unit and accepting responsibility.		

TASK STEPS AND PERFORMANCE MEASURES	GO	NO-GO
(2) As applicable, guides leading the relieving unit to specified positions in the determined sequence of relief, using covered and concealed routes and maintaining 360-degree security. ASSESS *8. Unit leaders, or designated representatives, report completion of the relief to the higher HQ commander. 9. The unit continues operations as directed. * indicates a leader task step.		

SUPPORTING INDIVIDUAL TASKS

Task Number	Task Title
071-410-0010	Conduct a Leader's Reconnaissance
071-326-5502	Issue a Fragmentary Order
071-331-0815	Practice Noise, Light, and Litter Discipline
071-326-5505	Issue an Operation Order at the Company, Platoon, or Squad Level
061-284-3040	Engage Targets with Close Air Support
061-283-6003	Adjust Indirect Fire

SUPPORTING COLLECTIVE TASKS

Task Number	Task Title
07-2-1342	Conduct Tactical Movement (Platoon-Company)
07-2-5036	Conduct Coordination (Platoon-Company)
07-2-6063	Maintain Operations Security (Platoon-Company)
07-2-9005	Conduct a Linkup (Platoon-Company)
07-2-9006	Conduct a Passage of Lines as the Passing Unit (Platoon-Company)
07-2-3036	Integrate Indirect Fire Support (Platoon-Company)

SUPPORTING BATTLE/CREW DRILLS

Drill Number	Drill Title
05-3-D0016	Conduct the 5 Cs
07-3-D9501	React to Contact (Visual, IED, Direct Fire [includes RPG])

TASK: Treat Casualties (08-2-0003)

(FM 4-25.11)   (AR 190-8)   (FM 4-02.7)

CONDITIONS: The unit has sustained casualties. The unit has medical treatment personnel and/or combat lifesavers. Threat force contact has been broken. Soldiers are wounded and may have chemical contamination or non-battle injuries. Unit personnel perform first aid (self-aid/buddy aid) treatment. The unit has analog and/or digital communications. A higher headquarters (HQ) operation order (OPORD) is available. Unit and higher HQ standing operating procedures (SOPs) are available. A treatment plan is available. This task is performed under all environmental conditions. The unit may be subject to attack by threat forces, including air; ground; chemical, biological, radiological, and nuclear (CBRN); or directed energy (DE) attack. Simplified collective protective equipment (SCPE) is on hand and/or field-expedient and natural shelters are available. Some iterations of this task should be performed in mission-oriented protective posture 4 (MOPP 4).

STANDARDS: Casualties are treated according to FM 4-25.11 and appropriate SOP(s). At MOPP 4 performance, degradation factors increase the time required to provide treatment and evacuation.

TASK STEPS AND PERFORMANCE MEASURES	GO	NO-GO

*1. The commander and leaders supervise first aid treatment of casualties (081-831-1055, 113-571-1022, 113-600-2001, 113-637-2001, and 805C-PAD-2060). They take the following actions:
    a. Implement treatment plan.
    b. Monitor treatment to ensure all casualties are treated.
    c. Direct employment of combat lifesavers to treat casualties.
    d. Monitor battlefield stress reduction and prevention procedures.
    e. Report casualties, as required.
    f. Coordinate with higher HQ for additional medical support.
    g. Coordinate replenishment of Class VIII supplies with supporting medical element according to SOPs.
    h. Direct distribution of Class VIII supplies according to SOPs.		

TASK STEPS AND PERFORMANCE MEASURES	GO	NO-GO
i. Enforce quality control procedures for Class VIII items issued to unit elements.		
2. Unit personnel perform first aid treatment (081-831-1003, 081-831-1005, 081-831-1007, 081-831-1008, 081-831-1025, 081-831-1026, 081-831-1032, 081-831-1033, 081-831-1034, 081-831-1044, 081-831-1045). They take the following actions:		
a. Evaluate casualties.		
b. Administer life-saving first aid treatment (cardiopulmonary resuscitation), if required.		
c. Control hemorrhage.		
d. Dress wounds.		
e. Splint suspected fractures.		
f. Provide first aid treatment to casualties with burns.		
g. Provide first aid treatment for environmental injuries.		
h. Provide first aid treatment for chemical casualties.		
i. Prevent shock.		
3. Unit medical personnel/combat lifesavers perform enhanced first aid treatment (081-831-0038, 081-831-0039, 081-831-1003, 081-831-1005, 081-831-1007, 081-831-1008, 081-831-1044, 081-831-1045, 081-833-0033, 081-833-0047, 081-833-0092). They take the following actions:		
a. Evaluate casualty for condition and type treatment needed.		
b. Measure casualty's vital signs.		
c. Initiate a field medical card.		
d. Insert oropharyngeal airway (J-Tube) in an unconscious casualty.		
e. Apply a splint to a fractured limb.		
f. Administer first aid to chemical agent casualties.		
g. Initiate an intravenous infusion for hypovolemic shock.		
h. Identify environmental injuries.		
i. Treat environmental injuries.		

TASK STEPS AND PERFORMANCE MEASURES	GO	NO-GO

j. Manage casualties with combat operational stress reactions.
  4. Unit medical personnel/combat lifesavers evacuate casualties to supporting medical element (081-831-0101, 081-831-1046, 081-833-0092). They take the following actions:
   a. Prepare casualties for evacuation.
   b. Identify litter team(s).
   c. Construct improvised litter from available material, as required.
   d. Secure casualty on litter.
   e. Employ appropriate manual carry if litter is not available.
   f. Transport casualty without causing further injury according to SOPs.
* indicates a leader task step.		

SUPPORTING INDIVIDUAL TASKS

Task Number	Task Title
081-831-0038	Treat a Casualty For a Heat Injury
081-831-0039	Treat a Casualty For a Cold Injury
081-831-0101	Request Medical Evacuation
081-831-1003	Perform First Aid to Clear an Object Stuck in the Throat of a Conscious Casualty
081-831-1005	Perform First Aid to Prevent or Control Shock
081-831-1007	Perform First Aid for Burns
081-831-1008	Perform First Aid for Heat Injuries
081-831-1025	Perform First Aid for an Open Abdominal Wound
081-831-1026	Perform First Aid for an Open Chest Wound
081-831-1032	Perform First Aid for Bleeding of an Extremity
081-831-1033	Perform First Aid for an Open Head Wound
081-831-1034	Perform First Aid for a Suspected Fracture
081-831-1044	Perform First Aid for Nerve Agent Injury
081-831-1045	Perform First Aid for Cold Injuries
081-831-1046	Transport a Casualty
081-831-1055	Ensure Unit Combat Lifesaver Requirements Are Met
081-833-0033	Initiate an Intravenous Infusion
081-833-0047	Initiate Treatment for Hypovolemic Shock
081-833-0092	Transport a Casualty with a Suspected Spinal Injury
113-571-1022	Perform Voice Communications

113-600-2001   Communicate Via a Tactical Telephone
113-637-2001   Communicate Via a Tactical Radio in a Secure Net
805C-PAD-2060 Report Casualties

SUPPORTING COLLECTIVE TASKS

Task Number	Task Title
07-2-3000	Conduct Support by Fire (Platoon-Company)
07-3-3027	Integrate Direct Fires (Platoon-Company)
07-2-3036	Integrate Indirect Fire Support (Platoon-Company)
07-2-5009	Conduct a Rehearsal (Platoon-Company)
07-2-5027	Conduct Consolidation and Reorganization (Platoon-Company)
07-2-5063	Conduct Composite Risk Management (Platoon-Company)
07-2-6063	Maintain Operations Security (Platoon-Company)
07-2-9006	Conduct a Passage of Lines as the Passing Unit (Platoon-Company)
07-3-9013	Conduct an Action on Contact
07-3-9017	Conduct Actions at Danger Areas
08-2-0004	Evacuate Casualties
44-3-3220	Perform Passive Air Defense Measures
44-3-3221	Perform Active Air Defense Measures

TASK: Evacuate Casualties (08-2-0004)

(FM 4-25.11) (AR 190-8)   (AR 385-10)   (AR 600-8-1)   (ATTP 4-02) (FM 4-02.7) (TC 3-34.489)

CONDITIONS: Unit personnel are wounded and some may be chemically contaminated. Threat force contact has been broken. Unit defenses are reorganized and established. Casualties are evacuated from defensive positions to designated casualty collection points. Wounded enemy prisoners of war (EPW) casualties are evacuated to designated casualty collection points (CCPs) with appropriate security. The unit has analog and/or digital communications. Higher headquarters (HQ) operation order (OPORD) is available. Unit and higher HQ standing operating procedures (SOPs) are available. This task is performed under all environmental conditions. The unit may be subject to attack by threat forces, to include air; ground; chemical, biological, radiological, and nuclear (CBRN); or directed energy (DE) attack. Simplified collective protective equipment (SCPE) is on hand and/or field-expedient and natural shelters are available. Some iterations of this task should be performed in mission-oriented protective posture 4 (MOPP 4).

STANDARDS: Casualties are evacuated as soon as tactical situation permitted in according to FM 4-25.11, OPORD, appropriate SOPs, and provisions of the Geneva Conventions. At MOPP 4, performance degradation factors increase the time required to evacuate casualties.

TASK STEPS AND PERFORMANCE MEASURES	GO	NO-GO

*1. The commander and leaders supervise evacuation of casualties (113-571-1022, 113-600-2001, 113-637-2001). They take the following actions:
    a.   Monitor casualty evacuation operations for compliance with SOPs.
    b.   Identify casualty collection points.
    c.   Identify evacuation requirements.
    d.   Supervise preparation of casualties for evacuation.
    e.   Coordinate evacuation of casualties from unit area with the area defense command post (CP) according to SOPs.
    f.   Coordinate security requirements for the pick-up site with subelements and area defense CP.
    g. Disseminate evacuation information to unit personnel.		

TASK STEPS AND PERFORMANCE MEASURES	GO	NO-GO

h.   Forward casualty feeder report and witness statements to the area defense CP according to SOPs.
2.   Unit personnel prepare casualties for evacuation (101-92Y-0005, 113-571-1022, 113-600-2001, 113-637-2001, 805C-PAD-2060). They take the following actions:
    a.   Provide first aid treatment to casualties (08-2-0003).
    b.   Report casualties, as required.
    c.   Collect classified documents such as signal operation instructions/signal supplemental instructions (SOI/SSI), maps, overlays, and key lists.
    d.   Secure custody of organizational equipment according to SOPs.
    e.   Forward casualty feeder reports to unit HQ according to SOPs.
3.   Unit personnel evacuate casualties to casualty collection points using manual carries (081-831-1046). They take the following actions:
    a.   Select type of manual carry appropriate to situation and injury.
    b.   Evacuate casualty without causing further injury.
4.   Unit personnel evacuate casualties to casualty collection points using litter carries (081-831-1046). They take the following actions:
    a.   Identify litter team(s).
    b.   Construct improvised litter from available material, as required.
    c.   Secure casualty on litter.
    d.   Evacuate casualty without causing further injury.
5.   Unit personnel evacuate casualties to a medical treatment facility (MTF) using available vehicles (081-831-1046). They take the following actions:
    a.   Load maximum number of casualties.
    b.   Secure casualties in vehicle.
    c.   Evacuate casualties without causing further injury.
*6. The commander and leaders request aeromedical evacuation (081-831-0101, 113-571-1022, 113-600-2001, 113-637-2001, 301-371-1050). They take the following actions:
    a. Transmit request according to OPORD and SOPs.		

TASK STEPS AND PERFORMANCE MEASURES	GO	NO-GO
b. Select landing site, which provides sufficient space for helicopter hover, landing, and take-off.		
c. Supervise removal of all dangerous objects likely to be blown about before aircraft arrival.		
d. Supervise security of landing site according to the SOPs.		
e. Ensure landing zone (LZ) is appropriately marked (light sets, smoke, and so forth) according to SOPs, if required.		
7. Unit personnel assist in loading ambulance (081-831-1046). They take the following actions:		
a. Employ proper carrying and loading techniques.		
b. Load casualties in the sequence directed by crew.		
c. Load casualties without causing unnecessary discomfort.		
d. Employ safety procedures according to SOPs.		
e. Employ environmental protection procedures according to SOPs.		
8. Unit personnel evacuate chemically contaminated casualties (031-503-1035, 081-831-1046). They take the following actions:		
a. Assume MOPP 4.		
b. Mark contaminated casualties according to SOPs.		
c. Notify supporting MTF that contaminated casualties are en route to their location.		
d. Evacuate casualties directly to a designated decontamination and treatment station.		
e. Protect casualties from further contamination during evacuation.		
9. Unit personnel evacuate EPW casualties (081-831-1046, 181-105-1001). They take the following actions:		
a. Maintain security of EPW casualties according to SOPs.		
b. Search EPW casualties for weapons and ordnance before evacuation.		
c. Evacuate EPW casualties according to the provisions of the Geneva Conventions and SOPs.		
* indicates a leader task step		

SUPPORTING INDIVIDUAL TASKS

Task Number	Task Title
031-503-1035	Protect Yourself From Chemical and Biological (CB) Contamination Using Your Assigned Protective Mask
081-831-0101	Request Medical Evacuation
081-831-1046	Transport a Casualty
101-92Y-0005	Enforce Compliance With Property Accountability Policies
113-571-1022	Perform Voice Communications
113-600-2001	Communicate Via a Tactical Telephone
113-637-2001	Communicate Via a Tactical Radio in a Secure Net
181-105-1001	Comply With the Law of War and the Geneva and Hague Conventions
301-371-1050	Implement Operations Security (OPSEC) Measures
805C-PAD-2060	Report Casualties

SUPPORTING COLLECTIVE TASKS

Task Number	Task Title
08-2-0003	Treat Casualties
08-2-0001	Conduct Battlefield Stress Reduction and Prevention Procedures

SUPPORTING BATTLE/CREW DRILLS

Drill Number	Drill Title
07-3-D9507	Evacuate a Casualty (Dismounted and Mounted)

TASK: Conduct a Screen (Platoon-Company) (17-2-9225)

(FM 3-20.971)   (FM 3-20.96)   (FM 3-20.98)

CONDITIONS: The unit conducts operations as part of a higher headquarters (HQ) and receives an operation order (OPORD) or fragmentary order (FRAGO) directing it to conduct a stationary or moving screen mission for a larger force. The order designates the general trace of the screen, the duration of the screen, and the time it must be established. Indirect fire is available. The unit has communications with higher, adjacent, and subordinate elements. The unit has been provided guidance on the rules of engagement (ROE). Coalition forces and noncombatants may be present in the operational environment. Some iterations of this task should be conducted during limited visibility conditions. Some iterations of this task should be performed in mission-oriented protective posture 4 (MOPP4).

STANDARDS: The unit conducts the screen according to unit standing operating procedures (SOPs), the order, and/or higher commander's guidance. The unit does not allow any enemy ground element to pass through the screen undetected and unreported. The unit maintains continuous surveillance of enemy reconnaissance and main body avenues of approach, detects all enemy activity in the area of operations (AOs), provides early warning of enemy approach to the screened force, and destroys or repels enemy reconnaissance elements within its capabilities until displacement criteria are met as specified in the operations order. The unit complies with the ROE.

TASK STEPS AND PERFORMANCE MEASURES	GO	NO-GO
PLAN		
*1. Unit leaders gain and/or maintain situational understanding (SU) using available communications equipment, maps, intelligence summaries, situation reports (SITREPs), and other available information sources. Intelligence sources include company intelligence support team (CoIST), human intelligence (HUMINT), signal intelligence (SIGINT), and imagery intelligence (IMINT) to include unmanned aircraft systems (UAS), and unattended ground sensors (UGSs).		
*2. Unit leaders receive an OPORD or FRAGO and issue a warning order (WARNO) to include at a minimum—		
a. The mission or nature of the screen.		
b. The time and place for issuing the OPORD.		

TASK STEPS AND PERFORMANCE MEASURES	GO	NO-GO

c. Units or elements participating in the screen.
   d. Specific tasks not addressed by unit SOPs.
   e. The timeline for the screen.
*3. The leader conducts troop-leading procedures (TLPs) to develop the order.
*4. Unit leaders coordinate with the protected force leader. They take the following actions:
   a. Coordinate any reinforcements necessary to accomplish the screen mission in depth.
   b. Coordinate the general trace of the screen and effective time, if appropriate.
   c. Reaffirm area of responsibility (AOR).
   d. Determine the interval to be maintained between the unit and the protected force.
   e. Determine battle/target handover criteria and graphic control measures.
   f. Coordinate special requirements or constraints, such as observing named areas of interest (NAIs) or target areas of interest (TAIs).
*5. Unit leaders plan the screen mission. They take the following actions:
   a. Coordinate with higher HQ and adjacent units to obtain required intelligence products, and initiating a terrain analysis using maps and other terrain products.
   b. Conduct a map reconnaissance. Take the following actions:
     (1) Identify screen trace, orientation, lateral and rear boundaries, and NAIs.
     (2) Identify enemy avenues of approach and possible objectives for enemy reconnaissance and infiltrating elements.
     (3) Identify and mark tentative control measures and dismount and remount points.
     (4) Select routes or sectors to facilitate rearward displacement.
     (5) Disseminate data to subordinate units via digital and/or conventional means.
   c. Integrate the fundamentals of security. Take the following actions:
     (1) Orient on the force, area, or facility to be protected.		

TASK STEPS AND PERFORMANCE MEASURES	GO	NO-GO
(2) Perform continuous reconnaissance.		
(3) Provide early and accurate warning.		
(4) Provide reaction time and maneuver space.		
(5) Maintain enemy contact.		
d. Conduct mission analysis. Take the following actions:		
(1) Identify the limits of the AO and area of interest (AOI).		
(2) Determine location, orientation, type, depth, and composition of obstacles.		
(3) Evaluate the enemy by considering the following information:		
(4) Determine tempo of the operation.		
(5) Identify the focus of the operation.		
(6) Determine enemy courses of action (ECOA).		
e. Develop the surveillance and reconnaissance plan. Take the following actions:		
(1) Develop a plan that answers the commander's information requirements (CCIRs)/priority intelligence requirements (IRs/PIRs) and accomplish his intent.		
(2) Integrate air reconnaissance assets and UASs, if available, forward of the screen line.		
f. Organize the unit to best accomplish the mission. Take the following actions:		
(1) Assign units to observe, identify, and report enemy actions.		
(2) Coordinate for additional combat and sustainment augmentation, as required.		
(3) Employ attached sustainment/protection elements, such as engineers, to provide support to maneuver elements.		
(4) Designate security forces to cover likely enemy approaches.		
g. Plan for air and ground integration.		
h. Plan for positioning of leadership elements.		
i. Designate which unit has responsibility for the area between the screening force rear boundary and the screened force AO.		

TASK STEPS AND PERFORMANCE MEASURES	GO	NO-GO
j. Designate graphic control measures, including—		
(1) Initial screen line that is forward of the general trace but within range of supporting indirect fire.		
(2) Subsequent screen lines as phase lines.		
(3) Passage of lines graphics and infiltration lanes.		
(4) Left and right limits of the screen as well as a phase line for the rear boundary.		
(5) Sectors, areas, or boundaries for subordinate elements.		
(6) Rally points, linkup points, contact points, and checkpoints.		
(7) General locations for observation posts (OP) enabling observation of the avenues of approach into the sector.		
k. Designate NAIs and assign observation.		
l. Annotate unit graphic control measures using higher HQ FRAGO overlay as a guide, and disseminating graphics to subordinate units.		
m. Establish engagement criteria according to—		
(1) Size of enemy force.		
(2) Type of enemy unit.		
(3) Activity of enemy unit.		
n. Integrate a fire support plan.		
o. Integrate the engineer obstacle plan.		
p. Plan sustainment and take the following actions:		
(1) Integrate the movement and positioning of sustainment assets into the scheme of maneuver.		
(2) Integrate refueling, rearming, and resupply operations into the scheme of maneuver.		
(3) Ensure adequate support to reconnaissance elements.		
(4) Plan supply routes to each element's location.		
(5) Plan immediate support to high-risk operations.		
(6) Plan and coordinate casualty evacuation assets.		
(7) Establish drop points for movement of key sustainment assets.		

TASK STEPS AND PERFORMANCE MEASURES	GO	NO-GO

(8) Plan on-order control measures, logistics release points (LRP), unit maintenance collection points (UMCP), and ambulance exchange points (AXPs).
    q. Develop contingency plan for chance contact with the enemy prior to reaching initial screen line.
    r. Plan movement of units performing front, rear, and/or flank screen.
    s. Plan limited visibility surveillance requirements.
    t. Plan successive bounds, alternate bounds by units, or continuous marching method of movement.
    u. Coordinate for passage of lines, if necessary.
    v. Ensure the plan is understood by all subordinate leaders.
PREPARE
*6. Unit leaders publish the order and distribute all paragraphs, annexes, and supporting overlays throughout the unit.
*7. Unit leaders give the order to execute screen.
EXECUTE
8. The unit moves to screen line by conducting one of the three primary methods:
    a. Zone reconnaissance to answer the intelligence requirement.
    b. Infiltration to avoid enemy forces and establishing the screen.
    c. A tactical road march to the screen line.
9. The unit occupies the screen. It takes the following actions:
    a. Establishes the stationary screen:
        (1) Determines changes to task organization and unit AOs after zone reconnaissance based on tasks and the factors of the mission given, enemy forces and their capabilities, terrain and weather effects, troops available, time available to execute the operation, and civil considerations (METT-TC).
        (2) Determines primary screen orientation for the unit and primary OPs.
        (3) Identifies engagement criteria.
        (4) Identifies method of displacement to subsequent screen lines while maintaining contact with the enemy.		

TASK STEPS AND PERFORMANCE MEASURES	GO	NO-GO
(5) Identifies initial locations for attached maneuver forces that provides flexible response against enemy reconnaissance throughout the unit AO if applicable.		
(6) Identifies positions that allow use of Long-Range Advanced Scout Surveillance System (LRAS3) to assist in observation and provides overwatch, if needed.		
(7) Determines requirements for short- and long-duration surveillance of NAIs.		
(8) Determines patrol requirements between or in support of OPs.		
b. Conducts reconnaissance that uses cueing, mixing, and redundancy to integrate unit and other assets into the security effort to gain and maintain contact throughout the depth of the AO. This should include taking the following actions:		
(1) Positions OPs in depth and focused on NAIs.		
(2) Employs UASs and aviation assets to reconnoiter routes, infiltration lanes, or key and restricted terrain forward or to the flanks of the unit AO.		
(3) Orients other surveillance and reconnaissance assets on NAIs located on avenues of approach, routes forward, or the flanks of the unit screen line.		
(4) Emplaces ground sensors on flank avenues of approach or routes leading into the unit AO.		
(5) Employs chemical, biological, radiological, and nuclear (CBRN) reconnaissance teams to reconnoiter templated attacks and bypasses. Takes the following actions:		
(a) Synchronizes target acquisition tasks with security and reconnaissance tasks		
(b) Synchronizes fires to suppress or destroy enemy elements or high-value targets (HVTs).		
(c) Establishes locations and criteria for RHO and target handover.		
(d) Specifies graphic control measures that support the concept of the operation.		
(e) Establishes moving screen. Takes the following actions:		

TASK STEPS AND PERFORMANCE MEASURES	GO	NO-GO
(1) Uses control measures to facilitate orientation of direction of movement and orientation of screen.		
(2) Repositions to stay oriented on the force it is screening. Takes the following actions:		
(a) Maintains continuous surveillance of unit AO. Takes the following actions:		
(b) Uses continuous marching when speed is required and contact is not likely.		
(c) Uses bounding by section or OPs, alternately or successively, when security is desired and contact is likely.		
(3) Conducts coordination with supporting air elements, as applicable.		
10. Acquires threat reconnaissance elements and destroys, if required, according to order. Takes the following actions:		
a. Coordinates with reconnaissance and surveillance assets and/or air elements, as applicable, to gain contact with enemy reconnaissance forward of the initial screen line and/or in restrictive terrain.		
b. Directs OPs and patrols to initially focus on reconnaissance avenues of approach as required.		
c. Uses indirect fires to impede and/or harass the threat according to fire support plan.		
e. Conducts counter reconnaissance to destroy, defeat, or repel all threat reconnaissance elements within capabilities and in accordance with engagement criteria.		
11. The unit gains contact with threat main body. It takes the following actions:		
a. Coordinates with reconnaissance and surveillance assets and/or air elements, as applicable, to gain and assist in maintaining contact with threat main body.		
b. Reorients OPs and patrols to focus on most likely avenues of approach and/or NAIs.		
c. Accepts reconnaissance handover/battle handover/target handover from reconnaissance and surveillance assets and/or air elements, as applicable.		
d. Acquires targets and executes indirect fires according to the fire support plan.		

TASK STEPS AND PERFORMANCE MEASURES	GO	NO-GO

e.   Continues operations as directed.
12. The unit displaces to the subsequent screen line. It takes the following actions:
    a.   Requests permission to displace.
    b.   Directs OPs facing most immediate threat to displace first.
    c.   Continues to adjust indirect fires.
    d.   Maintains contact with advancing threat elements.
    e.   Conducts reconnaissance handover/battle handover/target handover with other elements according to order and/or unit SOPs.
    f. Reports to higher HQ, as applicable, when set on the subsequent line.
    g.   Keeps higher HQ informed throughout the operation.
13. The unit completes the screen.
ASSESS
14. Unit leader maintains situational understanding and control conduct of the screen.
    a.   Direct elements to move, on order, to successive screen lines.
    b.   Use FRAGOs and graphic control measures to direct the moves.
    c.   Direct elements to report when they occupy new screen lines or OPs.
15. The unit consolidates and reorganizes as needed.
16. The unit continues operations as directed.
*Indicates a leader task step.		

SUPPORTING INDIVIDUAL TASKS

Task Number	Task Title
061-284-3040	Engage Targets with Close Air Support
171-620-0011	Conduct Zone-Area Reconnaissance at Company-Troop Level
171-123-4001	Prepare a Platoon Fire Plan
171-620-0061	Conduct a Screen at Company-Troop Level
171-620-0016	Conduct Route Reconnaissance at Company-Troop Level
171-620-0019	Plan Fire Support at Company-Troop Level
171-121-4051	Prepare a Situation Report (SITREP)

171-121-4046   Direct Emplacement and Activation of Early Warning
                 Systems
171-121-4004   Conduct a Screening Mission

SUPPORTING COLLECTIVE TASKS

Task Number   **Task Title**
07-3-9013   Conduct Action on Contact
17-2-4017   Conduct Target Acquisition (Platoon-Company)
17-2-9225   Conduct a Screen (Platoon0Company)
17-6-9225   Conduct a Screen (Battalion-Brigade)
17-3-2605   Conduct a Defense
07-2-5081   Conduct Troop-Leading Procedures (Platoon-Company)

SUPPORTING BATTLE/CREW DRILLS

Drill Number   **Drill Title**
07-3-D9504   React to Indirect Fire
05-3-D0016   Conduct the 5 Cs

TASK: Conduct Roadblock and Checkpoint Operations (19-3-2406)

(FM 3-39)   (FM 3-19.4)   (FM 5-0)

CONDITIONS: The element receives an order from higher headquarters (HQ) to establish a roadblock and/or checkpoint in its area of operations (AO). The local police or security forces may assist with the operations. The unit has received guidance on the rules of engagement (ROE), rules of interaction (ROI), and escalation of force (EOF). Some iterations of this task should be performed in mission-oriented protective posture 4 (MOPP 4).

STANDARDS: The element conducts roadblock and checkpoint operations. The element plans and constructs a roadblock and/or checkpoint according to the commander's guidance. The roadblock and/or checkpoint controls vehicular and pedestrian traffic by preventing passage or limiting entry to and exit from the specified area. The element complies with the ROE, ROI, and EOF, mission instructions, higher HQ order, and other special orders. The time required to perform this task is increased when conducting it in MOPP4.

TASK STEPS AND PERFORMANCE MEASURES	GO	NO-GO

*1. The element leader prepares to conduct roadblock and/or checkpoint operations by initiating troop-leading procedures. He takes the following actions:
 a. Conducts an estimate of the situation.
    (1) Conducts a detailed mission analysis.
    (2) Includes time for understanding and restating the mission.
    (3) Uses the backward-planning sequence to schedule troop-leading procedures.
    (4) Verifies the commander's critical information requirements.
    (5) Requests or conducts an intelligence preparation of the battlefield and a threat analysis.		

TASK STEPS AND PERFORMANCE MEASURES	GO	NO-GO
NOTE: Units should consider roadblocks and checkpoints as an information/intelligence source. In addition, the police intelligence operations (PIO) function represents military police (MP) capability to collect and process relevant information from many sources generally associated with policing activities or military police operations. As an integrating function, PIO describes an approach to all other MP functions that ensures their integration with all relevant police activities and organizations in the operations process and the AO.		
b. Issues a warning order to all squads.		
(1) Includes a mission statement (who, what, where, when, and why).		
(2) Includes friendly and hostile situations.		
(3) Includes general and special instructions.		
c. Makes a tentative plan.		
(1) Analyzes the mission using mission, enemy, terrain and weather, troops and support available, time available, civil considerations (METT-TC).		
(2) Compares courses of action.		
d. Coordinates with higher HQ for the eight-digit grid coordinates of the area to be used.		
e. Coordinates with adjacent and/or supported units in the element's AO.		
f. Implements plans that instruct the element to operate the roadblock and/or checkpoint for 24 hours continuously, if needed (based on the mission).		
g. Implements a standing operating procedure (SOPs) for moving the roadblock and/or checkpoint, as needed according to the security and operations plan.		
h. Disseminates and enforces the ROE, EOF guidelines, and ROI.		
(1) Ensures that personnel are aware of and follow the ROE, EOF, and ROI.		
(2) Ensures personnel know the rules regarding search, arrest, standoff distances, and the use of force.		
(3) Directs personnel to attack and disable all vehicles or personnel attempting to breach or flee.		
(4) Directs personnel to eliminate hostile elements and vehicles that initiate or return fire.		

TASK STEPS AND PERFORMANCE MEASURES	GO	NO-GO
(5) Directs personnel to eliminate hostile elements and vehicles that persist in attempting a breach. **NOTE:** Nonlethal tactics and capabilities are always supported, held in reserve, or overwatched by lethal capabilities.		
*2. The subordinate element leader prepares for the mission. He takes the following actions:		
a. Ensures that functionality checks are performed on communications equipment and digital systems, if available.		
b. Ensures that communications are established using Force XXI Battle Command-Brigade and Below (FBCB2), if available.		
c. Plans for a mission control cell to communicate with higher HQ and issue orders and/or reports (digitally), if available.		
d. Coordinates for mission essential logistical and sustainment supplies (such as Class I, II, III, IV, V supplies).		
e. Directs squad leaders to conduct a map reconnaissance and develop overlays with all known friendly forces and routes plotted by using maps and/or digital means.		
f. Determines the type of roadblock and/or checkpoint (deliberate or hasty) to establish, based on mission requirements.		
g. Briefs elements on the ROE, ROI, and EOF.		
h. Conducts a precombat inspection.		
i. Ensures that a sleep plan is established at the element level.		
j. Ensures that a medical evacuation plan is established.		
k. Ensures that military working dog (MWD) teams are available.		
l. Ensures that an explosive ordnance disposal (EOD) team is available.		

TASK STEPS AND PERFORMANCE MEASURES	GO	NO-GO
NOTE: In the event that the MWD detects explosives or other substances, the MWD team should withdraw immediately from the vehicle. The area should be evacuated and the chain of command notified. Radio operations are stopped until the MWD team and security forces reach a safe distance from the suspected explosives. The MWD team will stay behind a barrier at a safe distance in case the EOD team leader requests further detection assistance.		
m. Coordinates for interpreters, host nation police, or host nation authorities.		
n. Coordinates for logistical support for lighting when needed.		
o. Coordinates for engineer support for emplacing obstacles, barriers, and structures.		
*3. The subordinate element leader supervises the element and ensures that roadblocks and/or checkpoints are set up properly. He takes the following actions:		
a. Verifies the location and/or route of the roadblock and/or checkpoint with the element leader.		
(1) Ensures that the location provides good cover and concealment.		
(2) Ensures that the selected location is adequate for the creation of a roadblock and/or checkpoint.		
(3) Verifies the location with higher HQ.		
b. Establishes security and defensive positions.		
(1) Ensures that the roadblock and/or checkpoint is located at a defendable site.		
(2) Ensures that the roadblock and/or checkpoint has crew-served weapons that can provide overwatch and cover the entire site.		
(3) Directs the establishment of fields of fire that cover the approaches to the roadblock and/or checkpoint.		
(4) Ensures that the elements construct the roadblock and/or checkpoint according to the leader's guidance.		
*4. The subordinate element leader directs the element to prepare for roadblock and/or checkpoint operations. He takes the following actions:		
a. Identifies the exact area to set up the roadblock and/or checkpoint.		

TASK STEPS AND PERFORMANCE MEASURES	GO	NO-GO
b. Ensures that the roadblock and/or checkpoint cannot be seen from a distance. **NOTE:** This keeps drivers from turning off when they see the roadblock and/or checkpoint. If possible roadblock or checkpoint location should not allow for high speed approach. c. Ensures that the element uses existing culverts, bridges, deep cuts, sharp bends, or dips in the road to create a roadblock and/or checkpoint. d. Reports the exact location of the roadblock and/or checkpoint to the element leader. e. Ensures that fighting positions are prepared for each squad member. f. Establishes vehicle and personnel search procedures according to established authorizations, the SOP, and ROE. g. Assigns personnel to establish male and female search teams. h. Ensures that the assault force is in place to pursue those who attempt to avoid the roadblock and/or checkpoint. i. Ensures that the element marks all perimeter barriers, wires, and limits with warning signs, to include speed limit. **NOTE:** Warning signs should be posted in the native and English languages in the roadblock and/or checkpoint area. If possible warning signs should be placed at various distances leading up to RB/CP (Example: 150, 100, 50 meters, and STOP). Signs should also specify when deadly force is authorized for failure to comply with posted warnings (based on the established SOP, orders, ROE, and EOF). 5. The element constructs a roadblock. It takes the following actions: a. Positions the roadblock at or near an intersection or near an area that allows for vehicles to be easily rerouted or turned around. b. Positions the roadblock so that it does not allow unauthorized vehicles or enemy personnel to bypass. c. Places barricades along the road, shoulders, and ditches to channel passing traffic.		

TASK STEPS AND PERFORMANCE MEASURES	GO	NO-GO

d.   Ensures that there is adequate lighting for drivers to see the roadblock.
    e.   Positions squad vehicles in a covered and concealed location near the squad's position.
    f.   Ensures that if barriers are used across the roadway, they have an opening where slow-moving vehicles can enter to allow the search teams time to observe them closely.
    g.   Establishes security force positions.
    h.   Ensures that a translator is present at the roadblock.
  6. The element constructs a checkpoint. It takes the following actions:
    a.   Establishes a deliberate or hasty checkpoint.
NOTE: Establish a deliberate checkpoint when it will be in operation for a long period of time (13 hours or more). A deliberate checkpoint is permanent or semi-permanent and is typically constructed to protect an operating base or well-established main supply route.
    b.   Establishes a hasty checkpoint when used for a set period of time, usually a short duration (approximately 5 to 30 minutes).
    c.   Establishes security and defensive positions.
    d.   Establishes a checkpoint where it is hidden from distant view (usually for a hasty checkpoint).
    e.   Creates approach lanes that force traffic to slow down, and directs vehicles to the designated areas.
NOTE: Engineer tape, debris, trees, and rocks can be used for hasty checkpoints. Deliberate checkpoints should be constructed of more permanent structures that may require engineer support (such as barriers, dragon's teeth, concertina wire, caltrops, cement blocks, and buildings).
    f.   Establishes holding areas.
    g.   Establishes an initial search area or zone.
    h.   Establishes detailed search areas for personnel (male and female) and vehicles.
    i.   Establishes security force positions.
    j.   Ensures that there is adequate lighting for night operations.
    k. Ensures that a translator is present at the checkpoints.		

TASK STEPS AND PERFORMANCE MEASURES	GO	NO-GO

7. The element establishes vehicle checkpoint zones or areas. It takes the following actions:
    a. Establishes the initial search zone.
 NOTE: The initial search zone is a distant visual search area where vehicles and personnel are ordered (by visual or audio means) to stop at a clearly marked point before they actually enter the checkpoint. Personnel and vehicles can be visually searched from a predetermined distance (approximately 25 to 100 meters or as the mission dictates) while checkpoint operators remain behind a protective barrier or vehicle. Personnel are ordered to exit their vehicle, open their vehicle compartments, uncover or take out items from their vehicle, open or pull up their overgarments, turn around, and perform any other additional measures according to the SOP. This visual search is conducted before bringing personnel and vehicles into the checkpoint for a detailed search. Local support authorities can be used for this area. The initial search zone is more applicable to a deliberate checkpoint but can be used for a hasty checkpoint if the mission allows.
    b. Establishes a canalization zone.
 NOTE: Natural and/or artificial obstacles are used for a canalization zone to canalize vehicles into the checkpoint with no way to exit without the consent of personnel controlling the checkpoint. This zone encompasses the maximum effective range of the unit's weapons systems.
    c. Establishes a turning or deceleration zone.
 NOTE: The turning or deceleration zone forces vehicles to decelerate and make slow turns. If individuals attempt to maintain their speed they could crash into a series of obstacles.
    d. Ensures that fighting positions are prepared for each element member.
    e. Establishes a detailed search zone.
 NOTE: The detailed search zone is a relatively secure area where personnel and vehicles are positively identified and a complete detailed search is conducted. Blocking obstacles are used to isolate vehicles or individuals from others with overwatch protection from weapon positions. The search zone is further subdivided into three subordinate requirements.		

TASK STEPS AND PERFORMANCE MEASURES	GO	NO-GO
(1) Uses partitions or screened areas for privacy. (2) Provides all-around security, protective barriers, and rapid removal areas for personnel and detainees. (3) Ensures that the reaction force is located to respond to the checkpoint and provide immediate assistance (lethal and nonlethal) if required. f. Establishes a safe zone. **NOTE:** The safe zone is an assembly area for the checkpoint that allows personnel to rest, sleep, eat, and recover in relative security. Normally personnel should be rotated in and out of extended checkpoint operations but a safe zone is an essential requirement. 8. The element maintains security. It takes the following actions: a. Establishes a rest area for personnel near the search area so they can assemble quickly as a reserve force. b. Searches all vehicles and personnel for certain items (such as weapons, explosives, and contraband) as directed by the subordinate element leader. c. Positions sentries and patrols to prevent a possible ambush. d. Ensures that vehicle traffic, movement, and personnel are handled according to the established directives, SOP, and ROE. e. Stays alert to detect suspicious activity, vehicles, equipment, or personnel. 9. The element reacts to hostile actions. It takes the following actions: a. Fires warning shots (if authorized by the ROE) to deter the breach. b. Uses the minimum amount of force necessary to disarm infiltrating military or paramilitary forces. c. Attacks to disable all vehicles attempting to breach or flee. d. Destroys vehicles that initiate or return fires or persist in attempting a breach. 10. The element moves the roadblock and/or checkpoint to keep the enemy off balance (when applicable and the mission dictates). It takes the following actions:		

TASK STEPS AND PERFORMANCE MEASURES	GO	NO-GO
a. Follows the SOP for moving the roadblock and/or checkpoint.		
b. Keeps all unnecessary and prepackaged roadblock and/or checkpoint equipment on the vehicles.		
c. Conducts rehearsals for setting up, taking down, and moving the roadblock and/or checkpoint.		
d. Conducts a detailed brief with the oncoming shift, including all incidents that may occur during the shift.		
e. Improves the roadblock and/or checkpoint and security positions as time and the situation permit.		
*11. The subordinate element leader maintains contact with the elements to ensure that each roadblock and/or checkpoint is operating properly.		
*12. The element leader continues to monitor the execution of the operation and forwards information to the element headquarters. He takes the following actions:		
a. Compiles information into the platoon situation overlay or digital system if available.		
b. Forwards situation reports and spot reports to the higher HQ.		
*13. The element leader plans for follow-on and future missions.		
* indicates a leader task step.		

SUPPORTING INDIVIDUAL TASKS

Task Number	Task Title
171-121-4045	Conduct Troop-Leading Procedures
191-377-4254	Search a Detainee
191-377-4256	Guard Detainees

SUPPORTING COLLECTIVE TASKS

Task Number	Task Title
17-2-4017	Conduct Target Acquisition (Platoon-Company)
17-2-9225	Conduct a Screen (Platoon-Company)
17-3-2605	Conduct a Defense

SUPPORTING BATTLE/CREW DRILLS

Drill Number	Drill Title
19-4-D0105	Establish a Hasty Checkpoint
05-3-D0016	Conduct the 5 Cs

TASK: Conduct Logistics Package (LOGPAC) Support (63-2-4546)

(FM 4-0)   (FM 5-19)   (FM 55-30)

CONDITIONS: The unit receives an operations order (OPORD) and/or fragmentary order (FRAGO) to conduct resupply operations upon the arrival of the logistics package (LOGPAC), or the commander determines that routine or emergency resupply is necessary. The unit has established communications with subordinate, adjacent and higher headquarters (HQ), and is passing information according to the tactical standing operating procedure (TSOP). The unit has been provided guidance on the rules of engagement (ROE). Coalition forces and noncombatants may be present in the operational environment. This task is performed under all day and night environmental conditions. Threat capabilities cover a full spectrum to include information gathering; hostile force sympathizers; terrorist activities to include suicide bombings; and conventional, air supported, and reinforced squad operations in a chemical, biological, radiological, and nuclear (CBRN) environment. Some iterations of this task should be performed in mission-oriented protective procedure 4 (MOPP 4).

STANDARDS: The unit requests supplies/services necessary to restore it to fully mission capable (FMC) status. It receives supplies and services as available and conducts distribution as needed to subordinate elements. The unit completes resupply operations within the time specified in the OPORD and/or FRAGO, or command guidance. The unit complies with ROE. No friendly unit suffers casualties or equipment damage as a result of fratricide.

TASK STEPS AND PERFORMANCE MEASURES	GO	NO-GO

*1. The executive officer (XO)/first sergeant (1SG) monitors supply status and reports status as required by unit tactical standing operating procedure (TSOP). (101-92A-4216)
 	
a.   Compile accurate supply status (by class) from leaders of each platoon/section/element. Reports cover the following supply classes:
        (1) Class I (Rations).
        (2) Class II (Supplies and Equipment).
        (3) Class III (Petroleum, Oil, and Lubricants [POL] products).
        (4) Class IV (Construction/Barrier Materials).
        (5) Class V (Ammunition).
        (6) Class VI (Personnel Demand Items).
        (7) Class VII (Major End Items).
        (8) Class VIII (Medical Supplies).		

TASK STEPS AND PERFORMANCE MEASURES	GO	NO-GO
(9) Class IX (Repair Parts).		
(10)Class X (Nonmilitary Program Materials such as agriculture and economic development).		
(11)Water.		
b. Submit consolidated logistical status (LOGSTAT) report through unit commander to higher HQ S-4 and/or forward support company (FSC).		
2. Unit reports personnel status to the higher HQ S-1 using personnel status (PERSTAT) report, requests replacements, and processes reassignment/replacements.		
a. Platoon sergeants (PSGs) report personnel strength/losses (with battle roster numbers) to platoon/element leader and XO/1SG using PERSTAT.		
b. 1SG compiles report of personnel strength, losses, and battle roster changes and submits roll-up PERSTAT through the company commander to the higher HQ S-1.		
c. 1SG and PSGs reassign remaining personnel to ensure key positions are filled and critical weapons are manned.		
d. 1SG and PSGs assign replacements using the same criteria.		
e. Notifies s operations officer (SOO) when LOGPAC Operations vehicles are fully loaded and ready to move.		
f. Verifies that trail party is equipped to recover vehicles that develop maintenance problems during the combat resupply operations convoy.		
3. Unit reports vehicle status and requests resupply or other support as needed.		
a. PSGs and section leaders report vehicle and equipment status to include battle damage assessment (BDA), to platoon leaders and XO/1SG.		
b. PSGs and section leaders report maintenance, recovery, and evacuation support requirements to platoon leaders and XO/1SG.		
c. XO/1SG compiles platoon/section reports/requests and maintenance forecast and submits them to the higher HQ S-4 and/or supporting maintenance unit.		

TASK STEPS AND PERFORMANCE MEASURES	GO	NO-GO
d. They forward SP crossing report to HQ when unit elements have crossed the SP using FBCB2, MTS, or FM radio.		
e. They employ correct signal operating instructions/signal supplemental instructions (SOI/SSI) codes in all transmissions.		
f. They enforce march discipline using FBCB2, MTS, FM radio, or proper visual signals.		
*4. XO/1SG coordinate logistical package (LOGPAC) with higher HQ S-4 and/or forward support company (FSC) (191-379-4407). He takes the following actions:		
a. Verify status of resupply/support requests.		
b. Coordinate actions at the logistics release point (LRP).		
c. Assume position(s) along march route that provides command presence at points of decision for reaction to changing tactical situation.		
d. Maintain situational awareness at all times using FBCB2 and MTS.		
e. Forward en route CBRN information.		
f. Enforce all movement policies defined in the TSOP and movement order, with emphasis on formation, distances, speeds, passing procedures, and halts.		
g. Report all threat sightings using SALUTE (size activity location unit time equipment) Report format.		
h. Adjust formation distances and speed consistent with CBRN, terrain, and light conditions.		
i. Enforce security measures, with emphasis on air guards surveillance, manning of automatic weapons, and concealment of critical cargo.		
j. Inform vehicle operators by FBCB2, radio, MTS, or proper visual signals, any violations of march discipline, security procedures, or changes to established orders.		
k. Enforce communications security (COMSEC) measures to include radio silence periods according to the OPORD and SOI/SSI.		
5. The supply sergeant (under the supervision of the Headquarters and Headquarters Company [HHC] or FSC commander) assembles the LOGPAC. He takes the following actions:		

TASK STEPS AND PERFORMANCE MEASURES	GO	NO-GO
a. Obtain requested supplies from FSC or higher HQ S-4.		
b. Obtain Class II, IV, VI, and VII supplies from higher HQ S-4 personnel.		
c. Consolidate replacement personnel and those returning from medical treatment.		
d. Consolidate vehicles returning from maintenance.		
e. Obtain mail from higher HQ S-1.		
f. Obtain personnel action documents from S1 section (to include award, finance, and legal documents).		
*6. 1SG/XO meets LOGPAC elements at the LRP. He takes the following actions:		
a. Move to the LRP and meets the supply sergeant and LOGPAC.		
b. Supervise actions at LRP as coordinated and/or specified by unit SOP.		
c. Occupy hasty defensive positions with 360-degree protective coverage (passengers).		
d. Report scheduled halt to HQ.		
e. Direct performance of preventive maintenance checks and services (PMCS) on vehicles.		
f. Inspect vehicle loads for safety and security.		
g. Begin departure at time specified by orders or designated by platoon leader.		
h. Report resumption of march to headquarters.		
*7. 1SG/XO coordinates unit resupply. He takes the following actions:		
a. Determine method of resupply (service station or tailgate).		
b. Determine location(s) of resupply.		
c. Determine unit priority for resupply if all required supplies/services are not available.		
d. Determine unit order of resupply to include attachments.		
e. Execute LOGPAC operations according to TSOP or issues FRAGO notifying unit of changes to normal LOGPAC operations.		
f. Reports resumption of march to higher HQ.		
8. The unit receives service station resupply if applicable. The following actions are taken:		

TASK STEPS AND PERFORMANCE MEASURES	GO	NO-GO
a. 1SG/XO escort LOGPAC move to designated resupply location along covered and concealed route.		
b. The unit security element conducts link-up with 1SG/XO and LOGPAC to organize resupply site, establishing security and use available cover and concealment.		
c. 1SG/XO issues FRAGO to PSGs and section sergeants on the organization of the resupply site, specific locations of medics, maintenance, supply points, mortuary affairs collection points and enemy prisoners of war (EPW) collection points.		
d. Support platoons/sections/elements conduct tactical movement to resupply site.		
e. Support platoons/sections/elements conduct appropriate actions of service station resupply as directed by the commander and/or unit SOP.		
*9. Convoy commander conducts night convoy. He takes the following actions:		
a. Brief drivers on night conditions.		
b. Provide visual adjustment period if march began during daylight.		
c. Prepare vehicles for blackout conditions according to the TSOP.		
d. Maintain prescribed interval between vehicles.		
e. Direct the wearing of night vision goggles (selected personnel).		
f. Direct the wearing of regular eye protection goggles (all other personnel).		
g. Enforce the use of ground guides during poor visibility periods.		
*10. Convoy commander conducts convoy through an urban area. He takes the following actions:		
a. Verify all weight, height, and width restrictions along route of march.		
b. Employ close column formation.		
c. Ensure that vehicle drivers obey traffic control directions unless escorted by military or host nation (HN) police.		
d. Employ directional guide's at all critical intersections.		

TASK STEPS AND PERFORMANCE MEASURES	GO	NO-GO
*11. The convoy commander coordinates/monitors actions at the designated LRP. He takes the following actions: 　a.　Verify that lead vehicle has arrived at the LRP. 　b.　Verify that all vehicles have arrived at the LRP. 　c.　Release unit serials to the supported unit's 1SG or his/her designated represented representative. 　d.　Direct unit serial reassembly at the LRP following unit resupply actions. 　e.　Lead reassembled combat resupply operations convoy back to release point (RP) in the battalion field trains area. 　f.　Ensure that all back haul logistics commodities arrive at the proper location. 　g.　Forward situation report (SITREP) to headquarters using FBCB2, MTS, or radio. * indicates a leader task step.		

SUPPORTING INDIVIDUAL TASKS

Task Number	Task Title
101-92A-4216	Coordinate Logistical Requirements
191-379-4407	Plan Convoy Security Operations
101-92A-8030	Manage Unit Supply Operations

SUPPORTING COLLECTIVE TASKS

Task Number	Task Title
63-2-4519	Transport Supplies, Equipment, and Personnel
63-2-4000	Coordinate Replenishment/Sustainment Operations
07-2-5036	Conduct Coordination (Platoon-Company)

SUPPORTING BATTLE/CREW DRILLS

Drill Number	Drill Title
07-3-D9501	React to Contact (Visual, IED, Direct Fire [includes RPG])

TASK: Conduct Operational Decontamination (03-2-9224)

(FM 3-11.5)   (FM 3-11)

CONDITIONS: The element is operating in a contaminated environment. Performance degradation from mission-oriented protective posture 4 (MOPP4) is increasing and protective gear is in danger of contamination. The time and tactical situation permit the element to conduct operational decontamination. Replacement protective gear is available for each Soldier. For a nonsupported decontamination, decontamination equipment and supplies are available and operational. For a supported decontamination, an operational decontamination unit is available and is tasked to provide decontamination support. This task is always performed in MOPP4.

STANDARDS: The element decontaminates individual gear and conducts MOPP4 gear exchange (using the buddy team, triple team, or individual (emergency) method) without sustaining additional casualties from chemical, biological, radiological, and nuclear (CBRN) contamination. The element limits the contamination transfer hazard by removing gross chemical contamination from equipment. The element reduces radiological contamination to negligible risk levels according to the element's tactical standing operating procedure (TSOP) and field manual (FM) guidance and/or reduces chemical and biological (CB) contamination to accelerate the weathering process and eventually provide temporary relief from MOPP4.

TASK STEPS AND PERFORMANCE MEASURES	GO	NO-GO
*1. The element leader determines the extent of the contamination and establishes the priorities for decontamination. He takes the following actions: a. Receives input from subordinate leaders and staff. b. Directs decontamination priorities. 2. The element submits a request for decontamination to higher headquarters (HQ). The request should include, as a minimum, the following: **NOTE:** Decontamination operations should be done between one and six hours after becoming contaminated. a. The designation of the contaminated element. b. The location of the contaminated element. c. The frequency and call sign of the contaminated element.		

TASK STEPS AND PERFORMANCE MEASURES	GO	NO-GO
d. The time that the element became contaminated. e. The number of personnel requiring a MOPP gear exchange. f. The number of vehicles and equipment (by type) that are contaminated. g. The type of contamination. h. Special requirements (such as a patient decontamination station, recovery assets, and an element decontamination team). 3. The element coordinates with higher HQ. It takes the following actions: a. Obtains permission to conduct decontamination. b. Obtains the necessary support to conduct decontamination. c. Selects the link up point to meet supporting units (a company supply section, a company or battalion power-driven decontamination equipment [PDDE] crew, or a decontamination squad or platoon). d. Coordinates with supporting elements. e. Requests replacement MOPP gear. f. Coordinates with supporting units to determine if they need to exchange MOPP gear also. *4. The element leader and CBRN specialists select a site to conduct the operation and ensure that the selected site provides— a. Adequate overhead concealment. b. Good drainage. c. Easy access and exit routes (off the main routes). d. Close proximity to a water source large enough to support vehicle wash-down (plan for 100 gallons per vehicle). e. A large enough area to accommodate the elements involved in operational decontamination (110 square meters for both the vehicle wash-down site and the MOPP gear exchange site). 5. The element coordinates for operational decontamination support (a company or battalion PDDE crew or a decontamination unit). It takes the following actions: a. Notifies higher HQ of the site selected for the operational decontamination.		

TASK STEPS AND PERFORMANCE MEASURES	GO	NO-GO
b. Establishes communications with the decontamination unit.		
c. Ensures that the decontamination unit knows the link up locations and the selected decontamination site.		
6. The element and supporting units move to the decontamination site. They take the following actions:		
a. Meet at the link up point as coordinated.		
b. Provide security at the link up point and the decontamination site.		
7. The element prepares for operational decontamination. It takes the following actions:		
a. Sets up the decontamination site.		
(1) The supporting decontamination unit crew sets up a vehicle wash-down site.		
(2) The contaminated element sets up a MOPP gear exchange site no less than 50 meters upwind from the vehicle wash-down at a 45 degree angle.		
(3) The remainder of the element prepares its equipment for decontamination.		
b. Conducts preparatory actions in the predecontamination marshalling area.		
(1) Vehicle crews (except operators) dismount unless they have an operational overpressure system and an uncontaminated interior.		
(2) Dismounted crews remove mud and camouflage from vehicles.		
NOTE: The contaminated element provides personnel to do this when crews do not dismount.		
(3) Separated vehicles and dismounted crews—		
(a) Ensure that vehicle operators are briefed (include the use of overhead cover and concealment and proper intervals).		
(b) Ensure that vehicles are buttoned up (all doors, hatches, and other openings closed or covered to include muzzles).		
(4) Moves vehicles (with operators) to the vehicle wash-down site.		
(5) Moves dismounted crews and all other Soldiers in the contaminated element to the MOPP gear exchange site.		

TASK STEPS AND PERFORMANCE MEASURES	GO	NO-GO
*8. The noncommissioned officer in charge (NCOIC) of the decontamination unit supervises the operation of the vehicle wash-down site. He ensures that—		
a. Vehicle operators maintain proper intervals between vehicles while processing through the wash-down station.		
b. Decontamination crew washes vehicles properly.		
(1) Starts at the top and work down.		
(2) Sprays hot, soapy water for 2 to 4 minutes per vehicle.		
(3) Wears a toxicological agent-protective (TAP) or wet-weather gear over MOPP gear.		
(4) Monitors water consumption.		
c. Operators move to the MOPP gear exchange after vehicle has been washed down.		
d. Operators move to the assembly area (AA).		
9. The contaminated element conducts MOPP gear exchange. It takes the following actions:		
a. Prepares the equipment decontamination station with super tropical bleach (STB) dry mix.		
b. Briefs MOPP gear exchange participants on the procedures to be followed.		
c. Places the decontaminated individual equipment on a clean surface (such as plastic, a poncho, or similar material).		
d. Exchanges MOPP gear using the buddy team, triple team or individual (emergency) method.		
NOTE: The individual emergency method is used only when a person does not have a buddy to help and the risks of MOPP failure demands that an MOPP exchange occur.		
e. Moves to the AA after they complete the MOPP gear exchange.		
10. Supporting elements process through the MOPP gear exchange site.		
11. The supporting decontamination element cleans and marks the site and reports the area of contamination using a nuclear, biological, chemical CBRN 5 report to higher HQ.		

TASK STEPS AND PERFORMANCE MEASURES	GO	NO-GO

*12. The element leader accounts for all personnel and equipment after completing the operational decontamination.
*13. The element leader reports to higher HQ. He takes the following actions:
    a. Reports the completion of decontamination and the location of the vehicle wash-down and MOPP gear exchange decontamination sites.
    b. Requests permission to perform unmasking procedures if no hazards are detected through testing.
    c. Determines the adequacy of the decontamination and adjusts the MOPP level as required (after obtaining approval from higher HQ).
14. The element continues its mission.
* indicates a leader task step.		

SUPPORTING INDIVIDUAL TASKS

Task Number	Task Title
031-503-1019	React to Chemical or Biological (CB) Hazard/Attack
031-503-1021	Mark NBC Contaminated Area
031-503-1031	Use the Chemical Agent Monitor
031-503-1035	Protect Yourself from Chemical and Biological (CB) Contamination Using Your Assigned Protective Mask
031-503-1037	Detect Chemical Agents Using M8 or M9 Detector Paper
031-507-3014	Supervise Decontamination Procedures
113-571-1022	Perform Voice Communications
113-573-8006	Use an Automated Signal Operation Instruction (SOI)
551-721-1352	Perform Preventive Maintenance Checks

SUPPORTING COLLECTIVE TASKS

Task Number	Task Title
07-2-5009	Conduct a Rehearsal (Platoon-Company)
07-2-5063	Conduct Composite Risk Management (Platoon-Company)
07-2-5081	Conduct Troop-Leading Procedures (Platoon-Company)
07-2-6063	Maintain Operations Security (Platoon-Company)

SUPPORTING BATTLE/CREW DRILLS

Drill Number	Drill Title
07-3-D9483	React to Nuclear Attack
03-3-D0035	React to a Chemical Attack

TASK: Support Company Level Intelligence, Surveillance, and Reconnaissance (ISR) (34-5-0471)

(FM 2-19.4)   (FM 2-91.4)

CONDITION: The team is supporting a unit conducting stability operations or support operations in an operational environment. The company/troop area of operations (AO) and area of interest (AOI) are established. Command and Control (C2) Information Systems (INFOSYS) are operational and are passing information according to tactical standing operating procedures (TACSOP). Communications are established with the battalion S2 and adjacent teams for coordination of intelligence information, tasking, reporting, and collaboration. The team has digital analytical, biometric, cellular exploitation, document and media exploitation, and photographic tools available. The team has the rules of engagement (ROE) and the rules of interaction (ROI). Coalition forces, civilian noncombatants, governmental and nongovernmental organizations, and media organizations may be present in the operational environment. Some iterations of this task should be performed in mission-oriented protective posture 4 (MOPP4).

STANDARD: The team developed the unit intelligence, surveillance, and reconnaissance (ISR) plan, conducted patrol intelligence pre-briefings, facilitated walk-in informants, evaluated ISR reporting, and updated the unit ISR plan.

TASK STEPS AND PERFORMANCE MEASURES	GO	NO-GO
1. The team develops the company ISR plan.		
a. Determines priority intelligence requirement (PIR)/intelligence requirement (IR) that apply to the company AO and AOI.		
(1) Updates the situation map with current named areas of interest (NAI) and target areas of interest (TAI).		
(2) Obtains the company commander's initial PIR.		
(3) Recommends revised company PIR/IR to the company commander.		
(4) Receives the commander's approval of the PIR/IR.		
b. Determine which PIR and IR can be answered with organic assets taking into consideration availability, capability, sustainability, vulnerability, and performance history.		
c. Develop indicators and specific information requirements (SIRs) for each PIR.		

TASK STEPS AND PERFORMANCE MEASURES	GO	NO-GO
d. Review external assets tasked to perform collection against PIR/IR in the company AO and AOI.		
e. Develop ISR tasks for subordinate company elements.		
f. Develop an ISR synchronization matrix (ISM) depicting organic and external ISR collection to be performed in the company AO and AOI.		
g. Prepare the unit ISR plan.		
h. Disseminate ISR plan to company elements, higher S2, and adjacent ISTs.		
i. Submit requests for information (RFI) to the battalion S2 for external ISR collection against PIR/IR in the company AO and AOI.		
NOTE: Requests for external ISR collection should be put in terms of capabilities required, rather than specific types of ISR assets. Specific ISR assets may already be tasked and unavailable, whereas a request for a capability (for example, full motion video coverage of an NAI or MSR) could be tasked by higher headquarters to a number of potential ISR assets. The IST must also consider than many aerial assets require 72 hours prior notice for tasking thru the air tasking order (ATO).		
2. The team conducts patrol intelligence pre-briefings.		
NOTE: The company intelligence support team (CoIST) must have a standard patrol intelligence pre-brief format consistent with higher headquarters tasking and reporting requirements. The format should be included in the unit TACSOP. The patrol intelligence pre-brief is focused on providing information to the patrol, which is separate from the patrol mission briefing given by the small unit leader or commander leading the patrol.		
a. Provides updated information from intelligence preparation of the battlefield (IPB) products.		
(1) The effects of terrain and weather.		
(2) Updates to the company operational environment.		
(3) Description of the operational effects on the mission.		
(4) Evaluation of threat capabilities.		
(5) Assessment of threat courses of action.		

TASK STEPS AND PERFORMANCE MEASURES	GO	NO-GO
b. Describe significant activities in the company AO and AOI over that past 24-48 hours.		
c. Describe ISR collection assets and priorities.		
(1) Review the commander's PIR and IR.		
(2) Describe NAI within the company AO and AOI.		
(3) Describe specific expectations of SOR.		
(4) Provide the ISR matrix to the patrol.		
d. Provide updated graphics supporting the patrol mission.		
(1) Routes to be taken.		
(2) NAIs and TAIs relating to the patrol.		
(3) Location of objective(s) for the patrol.		
(4) Imagery of the patrol route, NAIs, and objectives.		
e. Provide current assessments and future expectations.		
f. Distribute the high-payoff target list (HPTL).		
(1) Describe each target on the list.		
(2) Distribute the HPTL to patrol members.		
g. Provide updates on key personalities in the company AO/AOI.		
(1) Spheres of influence (SOI).		
(2) Groups the person is associated with.		
(3) Events the person has participated in.		
(4) Threats posed by the individual.		
h. Distribute the be on the lookout (BOLO) list.		
i. Provide target packet folders for high-value targets (HVT) expected to be encountered during the patrol.		
j. Provide the patrol with automated tools for data collection, if available.		
(1) Biometric systems and updated database files.		
(2) Digital cameras.		
(3) Cellular exploitation (CELLEX) systems.		
(4) Electronic media exploitation (MEDEX) systems.		
(5) Evidence collection kits.		
(6) Blank target packet folders to be completed during tactical questioning or upon detaining an individual.		

TASK STEPS AND PERFORMANCE MEASURES	GO	NO-GO
k. Provide updates assessments on civilian considerations in terms of ASCOPE.		
l. Remind patrol members of the limitations to conducting tactical questioning.		
3. The team facilitates walk-in informants.		
a. Establishes a discreet location for the informant meeting.		
b. Coordinates security for the meeting location.		
c. Conducts screening of the walk-in informant.		
(1) Obtain informant identification data.		
(2) Allow the informant to convey his story.		
(3) Determine how the informant obtained the information.		
(4) Determine if the informant provided the information to anyone else.		
(5) Determine if the informant has reported previous information to U.S. forces.		
(6) Determine if the informant is willing to be re-contacted.		
d. Documents walk-in informant information.		
e. Notifies the commander and S2 of the informant's desire to talk.		
4. The team evaluates ISR reporting from unit elements.		
a. Determine relevancy of reporting to PIRs and IRs.		
b. Updates ISR synchronization matrix (ISM) as PIR and IR are answered.		
c. Provide feedback to collectors and exploiters.		
5. The team revises the unit ISR plan.		
a. Recommends to the commander the re-tasking of ISR assets.		
b. Updates the ISR plan.		
c. Disseminates the revised ISR plan to unit elements, the battalion S2, and adjacent ISTs.		

SUPPORTING INDIVIDUAL TASKS

Task Number   **Task Title**
052-192-3262   Prepare for an Improvised Explosive Device (IED) Threat Prior to Movement (Unclassified/For Official Use Only) (U//FOUO)

052-703-9107   Plan for an Improvised Explosive Device (IED) Threat in a COIN Environment (Unclassified/For Official Use Only) (U//FOUO)

052-703-9113   Plan for the Integration of C-IED Assets in a COIN Environment

052-703-9114   Respond to an IED at the Company Level

150-718-5315   Establish the Common Operational Picture

150-718-6717   Plan for Possible Improvised Explosive Device Threats

301-192-6001   Apply Predictive Analysis to Support Counter Improvised Explosive Device Operations

301-192-6002   Apply Pattern Analysis Products to Support Counter Improvised Explosive Device Operations

301-192-6003   Prepare Request for Intelligence, Surveillance, and Reconnaissance in Support of Counter Improvised Explosive Device Operations

SUPPORTING COLLECTIVE TASKS

Task Number   **Task Title**

34-5-0470   Provide Situational Awareness of the Company Area of Operations

34-5-0472   Provide Intelligence Support Team Input to Targeting

SUPPORTING BATTLE/CREW DRILLS

Drill Number   **Drill Title**

05-3-D0016   Conduct the 5 Cs

05-3-D0019   Conduct 5 and 25 Meter Checks

TASK: Provide Situational Awareness of the Company Area of Operations (34-5-0470)

(FM 2-01.3)   (FM 2-19.4)

CONDITIONS: The team is supporting a unit that is conducting stability operations or support operations in an operational environment. The company's area of operations (AO) and area of interest (AOI) are established. Command and control (C2) information systems (INFOSYS) are operational and are passing information according to tactical standing operating procedures (TACSOP). Communications are established with the battalion S2 and adjacent teams for coordination of intelligence information, tasking, reporting, and collaboration. The team has digital analytical, biometric, cellular exploitation, document and media exploitation, and photographic tools available. Coalition forces, civilian noncombatants, governmental and nongovernmental organizations, and media organizations may be present in the operational environment. Some iterations of this task should be performed in mission-oriented protective posture 4 (MOPP 4).

STANDARDS: The team provides situational awareness in terms of processing combat information, conducting patrol intelligence debriefings, processing information and material gathered during site exploitation, applying intelligence preparation of the battlefield (IPB) products at unit level, and presenting intelligence according to the TACSOP.

TASK STEPS AND PERFORMANCE MEASURES	GO	NO-GO
1. The team tracks significant activities that occur in the company AO and AOI. It takes the following actions:		
a. Logs events according to the TACSOP.		
b. Conducts event pattern analysis. Takes the following actions:		
(1) Ambushes.		
(2) Sniper incidents.		
(3) IED incidents.		
(4) Indirect fire incidents.		
(5) Murders.		
(6) Kidnappings.		
c. Incorporates the information into current databases or IPB products.		
d. Updates the situation map (SITMAP).		
2. The team updates pattern analysis products. It takes the following actions:		

TASK STEPS AND PERFORMANCE MEASURES	GO	NO-GO
a. Updates the coordinates register(s)/incident map of cumulative events occurring within the AO.		
b. Updates the pattern analysis plot sheet depicting the time and date of significant incidents occurring in the AO.		
3. The team updates link analysis products. It takes the following actions:		
a. Updates time event charts depicting events in chronological order.		
b. Updates association matrixes showing connectivity between key individuals and events or activity.		
c. Updates activities matrixes depicting an array of personalities compared against activities, locations, events, or other appropriate information.		
d. Updates link analysis diagrams depicting the connections between people, groups, or activities.		
4. The team conducts patrol intelligence debriefings. It takes the following actions:		
NOTE: The IST must have a standard patrol intelligence debrief format consistent with higher headquarters (HQ) reporting requirements. The format should be documented in the unit TACSOP.		
a. Obtains the patrol's observations of actions and inaction in named areas of interest (NAI). Takes the following actions:		
(1) Determines whether PIRs, IRs, and SORs have been answered.		
(2) Determines whether host nation information requirements have been answered.		
b. Collects target folders completed by the patrol.		
c. Obtains the following information about the route the patrol was tasked to take:		
(1) Status of the route.		
(2) Observations made along the route.		
d. Obtains the following patrol observations about the populace:		
(1) Key engagements with civilians during the patrol.		
(2) Topics discussed with members of the populace.		
(3) Observed or perceived attitudes of the populace.		

TASK STEPS AND PERFORMANCE MEASURES	GO	NO-GO
(4) Unusual activity among the population.		
(5) Unusual sights, sounds, or odors noticed by patrol members.		
(6) Assessments, observations, and notes from key leader engagements (KLE) during the patrol.		
(7) New posters, graffiti, or propaganda visible.		
e. Obtains changes to the terrain or physical environment in the AO or AOI.		
f. Obtains the patrol's town/village assessment.		
g. Obtains the patrol's host nation security force assessment.		
h. Obtains digital photographs made during the patrol. Takes the following actions:		
(1) Allows patrol members to describe what is occurring in each photo.		
(2) Obtains from patrol members the names of individuals in the photographs, if known.		
i. Enters all patrol data into available databases.		
j. Submits reports on patrol activities in accordance with the TACSOP.		
5. The team processes information obtained during unit site exploitation missions. It takes the following actions:		
a. Collects photographs taken during the mission.		
b. Obtains information gathered through tactical questioning during the site exploitation.		
c. Collects target packets completed during the site exploitation.		
d. Down loads biometric data collected.		
e. Collects data obtained through cellular exploitation (CELLEX) at the site.		
f. Collects data obtained through electronic media exploitation (MEDEX).		
g. Collects documents seized at the site.		
h. Collects all biometric, cellular exploitation, and media exploitation equipment from the element that conducted the site exploitation.		
i. Conducts analysis of all data collected to update company target folders.		
j. Forwards all collected material to higher echelon intelligence elements in accordance with the TACSOP.		
k. Updates databases according to the TACSOP.		

TASK STEPS AND PERFORMANCE MEASURES	GO	NO-GO
l. Reports site exploitation results to the battalion S2 and commander in accordance with the TACSOP.		
6. The team updates the analysis of the operational environment in the company AO and AOI. It takes the following actions:		
a. Analyzes the characteristics of the environment.		
b. Analyzes the military aspects of weather.		
c. Assesses the following civilian considerations (ASCOPE) in the operational environment:		
NOTE. The acronym ASCOPE refers to the civilian considerations of area, structures, capabilities, organizations, people, and events.		
(1) Areas included in the AO and AOI are the following:		
(a) Government centers.		
(b) Political boundaries.		
(c) Trade routes or main supply routes (MSR).		
(d) Commercial, market, and residential zones.		
(e) Social, political, religious, or criminal enclaves.		
(f) Agricultural and mining regions.		
(g) Displaced person or refugee centers.		
(2) Structures present in the AO and AOI are the following:		
(a) Street and urban patterns.		
(b) Power plants and dams.		
(c) Communications towers.		
(d) Religious buildings.		
(e) Television and radio stations.		
(f) Hospitals.		
(g) Subterranean routes within the area (tunnels and sewers).		
(h) Schools.		
(i) Prisons or jails.		
(3) Capabilities of public and commercial services in the AO and AOI are the following:		
(a) Law enforcement and fire services, to include their relationship to the military.		
(b) Electrical services.		
(c) Water supply and sewage.		
(d) Fuel distribution.		

TASK STEPS AND PERFORMANCE MEASURES	GO	NO-GO
(e) Transportation services. (f) Public communications. (g) Health services. (h) Availability of basic necessities such as food, clothing, and shelter. (4) Organizations that are factors in the AO and AOI are the following: (a) Host nation governmental agencies, to include military forces. (b) Religious groups or organizations. (c) Criminal organizations, to include their relationship to the population, political parties, and the police. (d) Labor organizations. (e) Community or fraternal organizations. (f) U.S. Government agencies. (g) Nongovernmental organizations (NGOs). (h) U.N. agencies. (5) People located within the AO and AOI are the following: (a) Demographics of the population to include such factors as ethnic groups, age distribution, and income groups. (b) Tribes and clans present. (c) Perceptions of the culture. (d) Loyalties of the population. (e) Authority figures present, to include village or tribal elders. (f) Key communicators. (6) Events within the AO and AOI are the following: (a) National or religious holidays. (b) Planned or anticipated civil disturbances. (c) Agricultural or marketplace cycles. (d) Elections. (e) Celebrations. d. Applies current rules of engagement (ROE) and legal restrictions (treaties or agreements) to civil considerations in the company AO and AOI. e. Analyzes the limits of the company AO and determines whether the AO is—		

TASK STEPS AND PERFORMANCE MEASURES	GO	NO-GO
(1) Consistent with specified plans and orders.		
(2) Sufficient to accomplish assigned missions.		
f. Analyzes the limits of the area of influence and the AOI. Takes the following actions:		
(1) Assesses the area of influence including all geographic areas, ethnically populated areas, religious factors, or economic factors that could impact company operations.		
(2) Determines whether the AOI is sufficient to include all areas from which the threat could impact company missions.		
g. Determines intelligence gaps in the aspects of the operational environment using existing databases.		
h. Initiates collection of information required to complete IPB. Takes the following actions:		
NOTE: Reconnaissance of the AO or AOI is offering the best means of confirming terrain analysis and answering gaps in knowledge of the military aspects of terrain.		
(1) Determines whether organic unit capabilities could answer gaps in information on the company's operational environment.		
(2) Recommends to the commander the use of organic unit ISR assets to fill intelligence gaps in the operational environment in the company AO.		
(3) Submits requests for information (RFI) to the battalion S2 for gaps the unit cannot answer.		
7. The team updates effects of the environment on company operations. It takes the following actions:		
a. Assesses the completeness of IPB terrain analysis as it relates to the company AO and AOI, including—		
(1) Cross country mobility.		
(2) Lines of communications (LOC) (transportation, communications, power).		
(3) Vegetation type and distribution.		
(4) Surface drainage and configuration.		
(5) Surface materials.		
(6) Obstacles.		
(7) Infrastructure.		
(8) Flood zones.		
(9) Rotary wing aircraft landing zones.		

TASK STEPS AND PERFORMANCE MEASURES	GO	NO-GO

b. Assesses the weather effects on terrain for suitable locations or routes in the company AO and AOI, including—
     (1) Observation posts.
     (2) Avenues of approach.
     (3) Infiltration and exfiltration routes.
     (4) Engagement areas.
     (5) Battle positions.
     (6) Collection asset or weapon system locations.
 c. Analyzes the military aspects of the terrain (OAKOC) in the company AO and AOI, including—
     (1) Observation and fields of fire.
     (2) Avenues of approach.
     (3) Key terrain.
     (4) Obstacles.
     (5) Cover and concealment.
 8.   The team updates the evaluation of the threat in the company AO and AOI. It takes the following actions:
 a. Analyzes threat capabilities, including—
     (1) Composition of threat forces and cells and their affiliated political, religious, or ethnic organizations.
     (2) Disposition of threat forces or cells within the company AO or AOI.
     (3) Threat force and cell tactics or accepted principles of operation.
     (4) Logistical or monetary support of the threat cells.
     (5) Operational effectiveness of the threat cells in the AO and AOI.
     (6) Level of training of threat cells.
     (7) Ability to recruit new personnel into threat cells.
     (8) Ability to travel for planning and coordination.
     (9) All forms of support available to threat cells, including—
         (a) Local support from the population.
         (b) Regional support in the form of sanctuary, security, or transportation.
         (c) National support that can be in the forms of moral, physical, or financial.		

TASK STEPS AND PERFORMANCE MEASURES	GO	NO-GO
b. Updates the threat model. Takes the following actions:		
(1) Determines whether pattern trends are consistent or changing.		
(2) Determines whether activities fit anticipated threat courses of action (COA).		
(3) Develops or updates threat tactics, techniques, and procedures (TTP).		
(4) Refines the threat situation template.		
(5) Forecasts threat future actions.		
(6) Identifies potential targets.		
(7) Converts threat patterns of operation to graphics.		
(8) Describes threat tactics and options.		
(9) Determines high-value targets (HVT) and high-payoff targets (HPT) located within the company AO and AOI.		
(10) Updates company named areas of interest (NAI).		
(11) Formulates proposed or updated company priority intelligence requirements (PIR).		
(12) Updates ASCOPE assessments.		
c. Assesses company operational trends for vulnerability to threat activities.		
d. Describes threat capabilities in terms of actions they can be expected to take.		
9. Analyzes threat courses of action in the company AO and AOI. Takes the following actions:		
a. Assesses the threats likely objectives and desired end state.		
b. Assesses the full set of courses of action available to the threat, including actions—		
(1) Disruptive to friendly force or coalition operations.		
(2) Likely to kill large numbers of friendly forces or civilians.		
(3) Effective in discouraging popular support for friendly forces.		
(4) Profitable in terms of amount of ransom gained.		
(5) Effective at gaining supporters.		
(6) With positive propaganda effect to be gained.		

TASK STEPS AND PERFORMANCE MEASURES	GO	NO-GO

c. Prioritizes threat courses of action in the company AO and AOI and determines the most—
     (1) Likely threat courses of action.
     (2) Dangerous threat courses of action.
 10. The team presents intelligence. It takes the following actions:
     a. Prepares the intelligence running estimate in accordance with the TACSOP.
     b. Disseminates the intelligence running estimate to the battalion S2 and adjacent units.
     c. Conducts an intelligence update briefing to the commander and key unit leaders.
 * indicates a leader task step.		

SUPPORTING INDIVIDUAL TASKS

Task Number	Task Title
052-192-3262	Prepare for an Improvised Explosive Device (IED) Threat Prior to Movement (Unclassified/For Official Use Only) (U///FOUO)
052-703-9107	Plan for an Improvised Explosive Device (IED) Threat in a COIN Environment (Unclassified/For Official Use Only) (U///FOUO)
052-703-9113	Plan for the Integration of C-IED Assets in a COIN Environment
052-703-9114	Respond to an IED at the Company Level
150-718-5315	Establish the Common Operational Picture
150-718-6717	Plan for Possible Improvised Explosive Device Threats
171-300-0083	Enforce Rules of Engagement (ROE)
301-192-6001	Apply Predictive Analysis to Support Counter Improvised Explosive Device Operations
301-192-6002	Apply Pattern Analysis Products to Support Counter Improvised Explosive Device Operations
301-192-6003	Prepare Request for Intelligence, Surveillance, and Reconnaissance in Support of Counter Improvised Explosive Device Operations

SUPPORTING COLLECTIVE TASKS

Task Number	Task Title
34-5-0471	Support Company Level Intelligence, Surveillance, and Reconnaissance (ISR)
34-5-0472	Provide Intelligence Support Team Input to Targeting

34-6-2039	Conduct Intelligence Preparation of the Battlefield (IPB) in Support of Urban Operations (BDE/BN)
34-6-2040	Conduct Intelligence Preparation of the Battlefield (IPB)
34-6-2041	Produce Intelligence Products

SUPPORTING BATTLE/CREW DRILLS

Drill Number	**Drill Title**
05-3-D0016	Conduct the 5 Cs
05-3-D0019	Conduct 4 and 25 Meter Checks

TASK: Provide Intelligence Support Team Input to Targeting (34-5-0472)

(FM 2-01.3)   (FM 2-19.4)

CONDITION: The team is supporting a unit conducting stability operations or support operations in an operational environment. The team is tasked to support to company efforts in the battalion targeting process. The company's area of operations (AO) and area of interest (AOI) are established. Command and Control (C2) Information Systems (INFOSYS) are operational and are passing information according to tactical standing operating procedures (TACSOP). Communications are established with the battalion S2 and adjacent teams for coordination of intelligence information, tasking, reporting, and collaboration. Coalition forces, civilian noncombatants, governmental and nongovernmental organizations, and media organizations may be present in the operational environment. Some iterations of this task should be performed in mission-oriented protective posture 4 (MOPP4).

STANDARD: The team maintained current high-payoff target lists (HPTLs) and high-value target lists (HVTLs). The team maintained updated target folders for high-value targets (HVTs) and high-value individuals (HVIs). The team provided input to company targeting priorities, pre-targeting meetings, and higher level targeting meetings.

TASK STEPS AND PERFORMANCE MEASURES	GO	NO-GO

1. The team maintains updated HPTLs and HVTLs.
     a. Obtains the current HPTLs and HVTLs from the battalion S2 Section.
     b. Identifies which targets from the HPTLs and HVTLs exist in the company area of operations.
     c. Recommend additional high payoff and high value targets specific to the company area of operations.
2. The team updates target folders for HVTs and HVIs.
     a. Obtains current target folders from the battalion S2.
     b. Updates target folder content based upon unit operations and activities.
          (1) Cover sheet information on the HVTs/HVIs.
               (a) Name.
               (b) Location.
               (c) Collection overview.
               (d) Photos.
               (e) Intelligence gaps on the HVTs/HVIs.
          (2) Physical description of the HVTs/HVIs.		

TASK STEPS AND PERFORMANCE MEASURES	GO	NO-GO
(a) Age.		
(b) Physical build.		
(c) Types of clothes worn.		
(d) Distinguishing physical features.		
(e) Height/weight.		
(f) Eye color.		
(g) Facial features.		
(h) Gait while walking.		
(i) Hair and the use of hats/headwear.		
(j) License plate numbers of all vehicles driven by HVTs/HVIs.		
(k) All vehicles used by the HVTs/HVIs.		
(l)All known aliases.		
(3) Background of the HVTs/HVIs.		
(a) Category in which the individual belongs (for example: jihadist).		
(b) Group affiliation.		
(c) Connections with government, military, or police.		
(d) Roles or functions the HVTs/HVIs provides.		
(e) The AO for the HVTs/HVIs.		
(f) Religious affiliation.		
(g) Province or region of origin.		
(h) Civilian education.		
(i) Military education.		
(j) Known disabilities.		
(k) Health status.		
(l) Travel patterns of the HVTs/HVIs.		
(m) Circumstances of any previous detentions.		
(n) Military or insurgency experience.		
(o) Expected actions, to include those if confronted.		
(4) List of actions or key events for which the HVTs/HVIs is responsible or involved.		
(5) HVTs/HVIs associates.		
(a) Photos of associates.		
(b) Associates descriptions.		
(c) Records of previous detentions.		
(d) Military or insurgency experiences.		

TASK STEPS AND PERFORMANCE MEASURES	GO	NO-GO

(e) Expected actions, to include those if confronted.
      (f) Aliases of all known associates.
    (6) HVTs/HVIs family.
      (a) Photos of family members.
      (b) Family background and descriptions.
      (c) Location(s) of family members.
      (d) Connections with government, military, or police.
      (e) Records of previous detentions.
      (f) Military or insurgency experience.
      (g) Expected actions, to include those if confronted.
      (h) Aliases of all family members.
   c. Includes HVTs/HVIs in relevant pattern and link analysis products.
  3. The team provides input into determining company level target priorities.
     a. Conducts analysis of intelligence, surveillance, and reconnaissance (ISR) collection in support of priority intelligence requirements (PIRs), specific information requirements (SIRs), and Specific orders and requests (SORs).
     (1) Includes combat information gathered from organic and nonorganic ISR assets.
     (2) Includes analysis of information from patrol debriefs.
     (3) Incorporates link and pattern analysis into target priority development.
     b. Categorize targeting priorities into logical target sets in support of operations.
 NOTE: The Battalion or BCT targeting cell or work group may have existing targeting sets in support of operations that the companies recommend input to.
     (1) Security of U.S. forces, coalition forces, or the population.
     (2) Governance capabilities of local, regional, or national government agencies.
     (3) Essential services available to the local population.
     c. Assists in determining desired targeting effects.		

TASK STEPS AND PERFORMANCE MEASURES	GO	NO-GO
(1) Recommends measures of performance for assets allocated to each target.		
(a) Outlines tasks to be performed by all subordinate elements during the targeting cycle.		
(b) Ensures that required actions are included in the target synchronization matrix.		
(2) Recommends measures of effectiveness for each target set.		
(a) Identify the desired end state or outcome of each target selected.		
(b) Recommends ISR collection to conduct combat assessment of the effectiveness of targeting.		
(c) Recommends whether the target requires follow-on lethal or nonlethal engagement.		
4. The team provides input to the unit pre-targeting meeting.		
a. Light and weather data provided by higher HQ.		
b. Terrain data in the form of maps or imagery.		
c. HVTLs with link and pattern analysis.		
d. Current intelligence requirements.		
(1) PIRs.		
(2) SIRs.		
(3) SORs.		
e. Threat courses of action (COA) and event template.		
f. Battalion ISR plan for the next 72 hours.		
g. Available ISR assets.		
(1) Organic unit assets.		
(2) Nonorganic assets.		
5. The team provides input to the unit targeting meeting.		
a. Light, weather, and terrain data.		
b. Current situational awareness products.		
(1) Situation template (SITEMP).		
(2) Incident overlays.		
(3) Link analysis products.		
(4) Pattern analysis products.		
c. Status off nonorganic ISR assets/capabilities requested by the company.		
d. Threat assessments.		
(1) Battle damage assessment of attacked targets during the past 12-24 hours.		

TASK STEPS AND PERFORMANCE MEASURES	GO	NO-GO

(2) Assessment of the effectiveness of non-kinetic targeting during the past 12-24 hours.
    (3) Changes to threat capabilities as a result of attacks.
  e. The status of—
    (1) Current and proposed PIRs.
    (2) Current and proposed SIRs.
    (3) Current and proposed SORs.
    (4) HVTs.
  f. Analysis of COAs for the targeting period.		

SUPPORTING INDIVIDUAL TASKS

Task Number	Task Title
052-703-9107	Plan for an Improvised Explosive Device (IED) Threat in a COIN Environment (Unclassified/For Official Use Only) (U//FOUO)
052-703-9113	Plan for the Integration of C-IED Assets in a COIN Environment
150-718-5315	Establish the Common Operational Picture
150-718-6717	Plan for Possible Improvised Explosive Device Threats
301-192-6001	Apply Predictive Analysis to Support Counter Improvised Explosive Device Operations
301-192-6002	Apply Pattern Analysis Products to Support Counter Improvised Explosive Device Operations
301-192-6003	Prepare Request for Intelligence, Surveillance, and Reconnaissance in Support of Counter Improvised Explosive Device Operations

SUPPORTING COLLECTIVE TASKS

Task Number	Task Title
34-5-0470	Provide Situational Awareness of the Company Area of Operations
34-5-0471	Support Company Level Intelligence, Surveillance, and Reconnaissance (ISR)

SUPPORTING BATTLE/CREW DRILLS

Drill Number	Drill Title
05-3-D0016	Conduct the 5 Cs
05-3-D0019	Conduct 5 and 25 Meter Checks

Chapter 3

Supporting Battle/Crew Drills

This chapter provides the company commander an example of the weapons and antiarmor company METL collective tasks listing with supporting battle and/or crew drills. This chapter also provides drill T&EOs that can be used to train or evaluate a single task drill. Several T&EOs may be used by an observer controller as an evaluation outline or by a commander as a training outline.

BATTLE/CREW DRILLS

3-1.   The METL collective task to drill table (see Table 3-1) is an example developed by the DOTD, MCoE. This table can be used by the company commander and unit leaders as an example to create their own unique unit METL to drill crosswalk.

3-2.   The drills shown in the example matrix found in Table 3-1 are displayed using the T&EO outline format. For more information on other drills the company may be expected to perform, see DTMS.

Table 3-1. METL collective task to drill table

METL Collective Task Number and Title	
Supporting Battle Drill Number and Title	
07-2-1090	**Conduct a Movement to Contact (Platoon-Company)**
	07-3-D9501, React to Contact (Visual, IED, Direct Fire [includes rocket propelled grenade [RPG])
	07-3-D9505, Break Contact
	17-3-D8008, React to an Obstacle
07-2-1256	**Conduct an Attack by Fire (Platoon-Company)**
	07-3-D9501, React to Contact (Visual, IED, Direct Fire [includes RPG])

Table 3-1. METL collective task to drill table (continued)

METL Collective task number and title	
Supporting Battle Drill Number and Title	
07-2-1252	**Conduct an Antiarmor Ambush (Platoon-Company)**
	07-3-D9508, Establish Security at the Halt
	07-3-D9501, React to Contact (Visual, IED, Direct Fire [includes RPG])
07-2-1324	**Conduct Area Security (Platoon-Company)**
	07-3-D9501, React to Contact (Visual, IED, Direct Fire [includes RPG])
	05-3-D0017, React to an IED Attack While Maintaining Movement
	19-4-D0105, Establish a Hasty Checkpoint
07-2-3000	**Conduct Support by Fire (Platoon-Company)**
	07-3-D9504, React to Indirect Fire
07-2-3036	**Integrate Indirect Fire Support (Platoon-Company)**
	07-3-D9508, Establish Security at the Halt
	07-3-D9501, React to Contact (Visual, IED, Direct Fire [includes RPG])
07-2-4054	**Secure Civilians During Operations (Platoon-Company)**
	05-3-D0016, Conduct the 5 Cs
07-2-5027	**Conduct Consolidation and Reorganization (Platoon-Company)**
	05-3-D0016, Conduct the 5 Cs
	07-3-D9507, Evacuate a Casualty (Dismounted and Mounted)
07-2-9001	**Conduct an Attack (Platoon-Company)**
	07-3-D9501, React to Contact (Visual, IED, Direct Fire [includes RPG])
	07-3-D9410, Enter a Trench to Secure a Foothold
	07-3-D9412, Breach of a Mined Wire Obstacle

Table 3-1. METL collective task to drill table (continued)

METL Collective Task Number and Title	
Supporting Battle Drill Number and Title	
07-2-9002	**Conduct a Bypass (Platoon-Company)**
	17-3-D8008, React to an Obstacle
	07-3-D9505, Break Contact
	05-3-D0016, Conduct the 5 Cs
07-2-9003	**Conduct a Defense (Platoon-Company)**
	07-3-D9501, React to Contact (Visual, IED, Direct Fire [includes RPG])
	17-3-D8004, React to Air Attack
07-2-9004	**Conduct a Delay (Platoon-Company)**
	07-3-D9501, React to Contact (Visual, IED, Direct Fire [includes RPG])
	07-3-D9505, Break Contact
	05-3-D0016, Conduct the 5 Cs

TASK: React to Contact (Visual, IED, Direct Fire [includes RPG]) (07-3-D9501)

CONDITIONS: Visual (dismounted/mounted). The unit is stationary or moves, conducting operations. Visual contact is made with the enemy. **Mounted.** The unit is stationary or moves, conducting operations. Visual contact is made with the enemy. Improvised explosive device (IED) (dismounted/mounted). The unit is stationary or moves, conducting operations. The unit identifies and confirms an IED or one is detonated. Direct fire dismounted/mounted. The unit is stationary or moves, conducting operations. The enemy initiates contact with a direct fire weapon.

CUE: This drill begins when visual contact, direct fire, or an IED is identified or detonated.

STANDARDS: Visual (dismounted). The unit destroys the enemy with a hasty ambush or an immediate assault through the enemy position. Visual (mounted). Based on the composition of the mounted unit, the unit either suppresses and reports the enemy position and continues its mission, or suppresses the enemy position for a follow-on assault to destroy them. IED (dismounted/mounted). The unit takes immediate action by using the 5 Cs procedure (confirm, clear, call, cordon, check, and control). Direct fire (dismounted/mounted). The unit immediately returns well-aimed fire and seeks cover. The unit leader reports the contact to higher headquarters (HQ).

TASK STEPS AND PERFORMANCE MEASURES

1. Visual dismounted.

    a. Hasty ambush. Unit leaders take the following actions:

       (1) Determine that the unit has not been seen by the enemy.

       (2) Signal Soldiers to occupy best available firing positions.

       (3) Initiate the ambush with the most casualty-producing weapon available, immediately followed by a sustained well-aimed volume of effective fire.

       (4) If the unit is prematurely detected, the Soldier(s) aware of the detection initiates the ambush.

       (5) Ensure the unit destroys the enemy or forces them to withdraw.

       (6) Report the contact to higher HQ.

    b. Immediate assault.

       (1) The unit and the enemy simultaneously detect each other at close range.

TASK STEPS AND PERFORMANCE MEASURES

(2) All soldiers who see the enemy engage and announce "contact" with a clock direction and distance to enemy, (example, "contact three o'clock, 100 meters"). Unit personnel take the following actions:

(3) Elements in contact immediately assault the enemy using fire and movement.

(4) The unit destroys the enemy or forces them to withdraw.

(5) The unit leader reports the contact to higher headquarters.

2. Visual mounted. Unit personnel take the following actions:

a. The Soldier who spots the enemy announces the contact.

b. The element in contact immediately suppresses the enemy.

c. The vehicle commander of the vehicle in contact sends contact report over the radio.

d. The unit maneuvers on the enemy or continues to move.

e. Vehicle gunners fix and suppress the enemy positions.

f. The unit leader reports the contact to higher HQ.

3. IED dismounted/mounted. Unit personnel take the following actions:

a. React to a suspected or known IED prior to detonation by using the 5 Cs.

b. Unit determines if there is a requirement for explosive ordnance disposal (EOD), while maintaining as safe a distance as possible and 360 security, Unit "confirms" the presence of an IED by using all available optics to identify any wires, antennas, detcord, or parts of exposed ordinance. Take the following actions:

(1) Conduct surveillance from a safe distance.

(2) Observe the immediate surroundings for suspicious activities.

(3) Requests EOD if the need is determined.

c. Unit "clears" all personnel from the area a safe distance to protect them from a potential second IED.

d. Unit "cordons" off the area, directs personnel out of the danger area, prevents all military or civilian traffic from passing and allows entry only to authorized personnel. They take the following actions:

(1) Direct people out of the 300-meter minimum danger area.

(2) Identify and clears an area for an incident control point (ICP).

(3) Occupy positions and continuously secure the area.

e. Unit "checks" the immediate area for secondary/tertiary devices around the incident control point (ICP) and cordon using the 5/25 meter checks.

f. Unit "controls" the area inside the cordon to ensure only authorized access.

g. Unit continuously scans the area for suspicious activity. They take the following actions:

TASK STEPS AND PERFORMANCE MEASURES

(1) Identify potential enemy observation, vantage, or ambush points.

(2) Maintain visual observation on the IED to ensure the device is not tampered with.

4. Direct fire dismounted. (See Figure 1.) Unit personnel take the following actions:

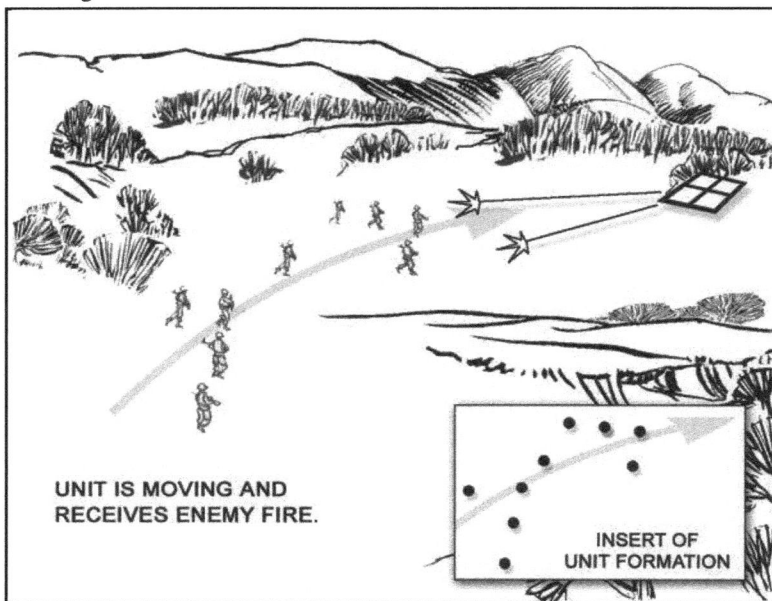

Figure 1. React to contact, direct fire (dismounted)

a. Soldiers under direct fire immediately return fire and seek the nearest covered positions. They call out distance and direction of direct fire. (See Figure 2.)

TASK STEPS AND PERFORMANCE MEASURES

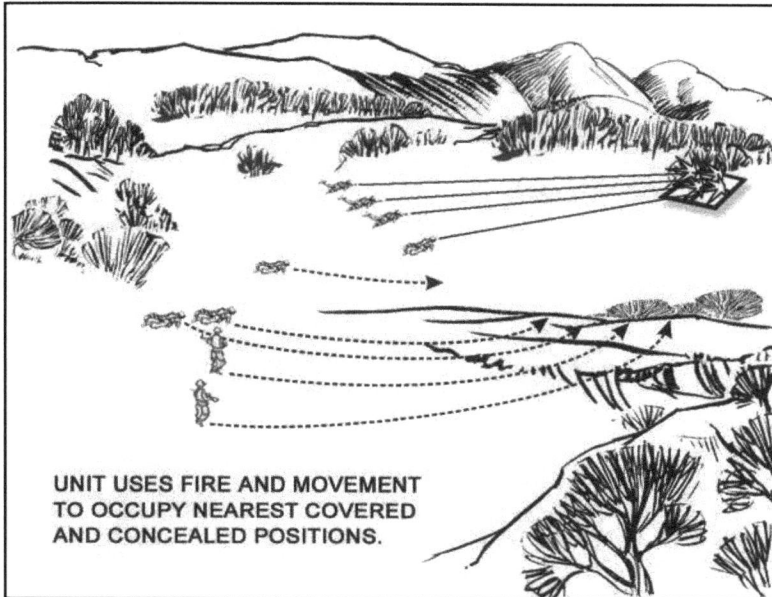

UNIT USES FIRE AND MOVEMENT
TO OCCUPY NEAREST COVERED
AND CONCEALED POSITIONS.

Figure 2. React to contact, direct fire (dismounted) (continued)

b. Element leaders locate and engage known or suspected enemy positions with well-aimed fire and pass information to the unit leader.

c. Element leaders control their Soldier's fire by—

(1) Marking targets with lasers.

(2) Marking the intended target with tracers or M203 rounds.

d. Soldiers maintain contact (visually or orally) with the Soldiers on their left or right.

e. Soldiers maintain contact with their team leader and relay the location of enemy positions. (See Figure 3.)

TASK STEPS AND PERFORMANCE MEASURES

UNIT IS IN POSITION ENGAGING
ENEMY WITH WELL-AIMED FIRES.

Figure 3. React to contact, direct fire (dismounted) (continued)

    f. Element leaders (visually or orally) check the status of their Soldiers.

    g. Element leaders maintain contact with the unit leader.

    h. Unit leader reports the contact to higher headquarters.

5.   Direct fire mounted. Unit personnel take the following actions:

    a. If moving as part of a logistics patrol, vehicle gunners immediately suppress enemy positions and continue to move.

    b. Vehicle commanders direct their drivers to accelerate safely through the engagement area.

    c. If moving as part of a combat patrol, vehicle gunners suppress and fix the enemy allowing others to maneuver against and destroy the enemy.

    d. Leaders (visually or orally) check the status of their Soldiers and vehicles.

    e. Unit leader reports the contact to higher HQ.

SUPPORTING PRODUCTS

Product ID	Product Name
FM 3-21.8	The Infantry Rifle Platoon and Squad
FM 3-21.75	Warrior Ethos and Soldier Combat Skills
ATTP 3-21.9	SBCT Infantry Rifle Platoon and Squad

SUPPORTING INDIVIDUAL TASKS

Task Number	Task Title
071-030-0004	Engage Targets with an MK 19 Grenade Machine Gun
071-054-0004	Engage Targets with an M136 Launcher
071-325-4407	Employ Hand Grenades
071-311-2130	Engage Targets with an M203 Grenade Launcher
071-010-0006	Engage Targets with an M249 Machine Gun
071-025-0007	Engage Targets with an M240B Machine Gun
071-100-0030	Engage Targets with an M16-Series Rifle/M4-Series Carbine
071-326-0502	Move Under Direct Fire
071-326-5611	Conduct the Maneuver of a Squad
071-326-5630	Conduct Movement Techniques by a Platoon
071-121-4080	Send a Spot Report (SPOTREP)
061-283-1011	Engage Targets with Indirect Fires
113-571-1022	Perform Voice Communications

SUPPORTED COLLECTIVE TASKS

Task Number	Task Title
07-2-1090	Conduct a Movement to Contact (Platoon-Company)
07-2-1450	Secure Routes (Platoon-Company)
07-2-9002	Conduct a Bypass (Platoon-Company)
07-2-9009	Conduct a Withdrawal (Platoon-Company)

TASK: Break Contact   (07-3-D9505)

CONDITIONS: (Dismounted/Mounted) - The unit is stationary or moving, conducting operations. All or part of the unit is receiving enemy direct fire.

CUE: The unit leader initiates drill by giving the order, BREAK CONTACT.

STANDARDS: (Dismounted/Mounted) - The unit returns fire. A leader identifies the enemy as a superior force, and makes the decision to break contact. The unit breaks contact using fire and movement. The unit continues to move until the enemy cannot observe or place effective fire on them. The unit leader reports the contact to higher headquarters (HQ).

TASK STEPS AND PERFORMANCE MEASURES
1. **Dismounted.**

a. The unit leader designates an element to suppress the enemy with direct fire as the base-of-fire element.

b. The unit leader orders distance, direction, a terrain feature, or last rally point for the movement of the first element.

c. The unit leader calls for and adjusts indirect fire to suppress the enemy positions.

d. The base-of-fire element continues to suppress the enemy. (See Figure 1.)

TASK STEPS AND PERFORMANCE MEASURES

UNIT IS ENGAGING ENEMY
AND MUST BREAK CONTACT.

Figure 1. Break contact (dismounted)

e. The bounding element uses the terrain and/or smoke to conceal its movement and bounds to an overwatch position.

f. The bounding element occupies their overwatch position and suppresses the enemy with "well-aimed fire." (See Figure 2.)

TASK STEPS AND PERFORMANCE MEASURES

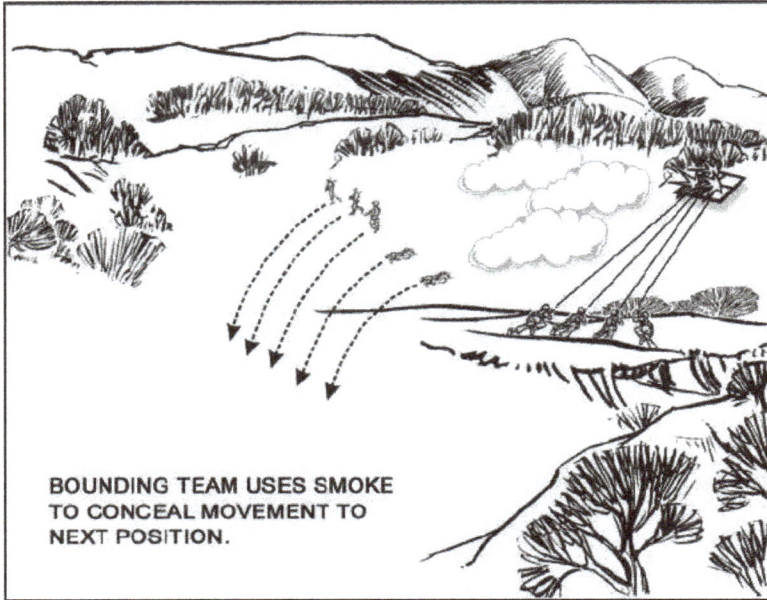

BOUNDING TEAM USES SMOKE
TO CONCEAL MOVEMENT TO
NEXT POSITION.

Figure 2. Break contact (dismounted) (continued)

g. The base-of-fire element moves to its next covered and concealed position. (Based on the terrain and volume and accuracy of the enemy's fire, the moving element may need to use fire and movement techniques). (See Figure 3.)

TASK STEPS AND PERFORMANCE MEASURES

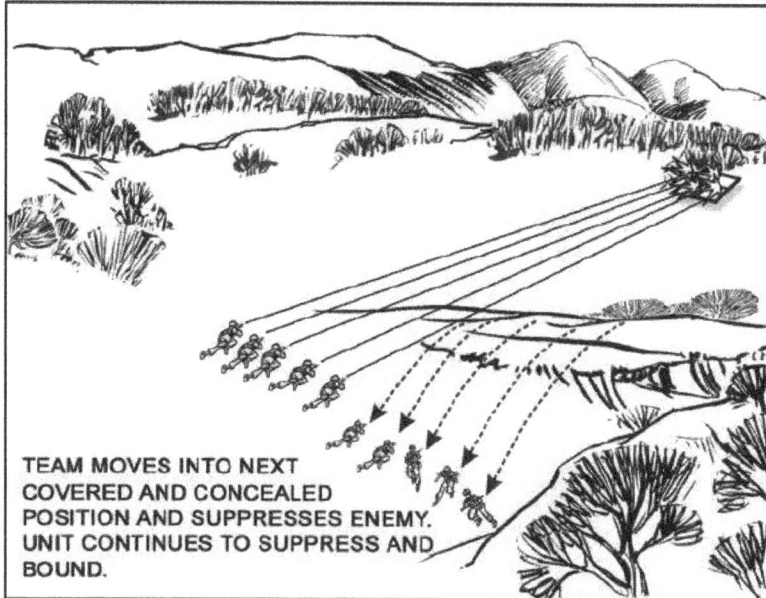

TEAM MOVES INTO NEXT
COVERED AND CONCEALED
POSITION AND SUPPRESSES ENEMY.
UNIT CONTINUES TO SUPPRESS AND
BOUND.

Figure 3. Break contact (dismounted) (continued)

h. The unit continues to suppress the enemy and bound until it is no longer in contact with enemy.

i. The unit leader reports the contact to higher headquarters.

2. **Mounted.**

a. The unit leader directs the vehicles in contact to place "well-aimed" suppressive fire on the enemy positions.

b. The unit leader orders distance, direction, a terrain feature, or last objective rally point over the radio for the movement of the first section.

c. The unit leader calls for and adjusts indirect fire to suppress the enemy positions.

d. Gunners in the base-of-fire vehicles continue to engage the enemy. They attempt to gain fire superiority to support the bound of the moving section.

e. The bounding section moves to assume the overwatch position.

(1) The section uses the terrain and/or smoke to mask movement.

(2) Vehicle gunners and mounted Soldiers continue to suppress the enemy.

f. The unit continues to suppress the enemy and bounds until it is no longer receiving enemy fire.

g. The unit leader reports the contact to higher HQ.

SUPPORTING PRODUCTS

Product ID	Product Name
FM 3-21.8	The Infantry Rifle Platoon and Squad
FM 3-21.75	Warrior Ethos and Soldier Combat Skills
ATTP 3-21.9	SBCT Infantry Rifle Platoon and Squad

SUPPORTING INDIVIDUAL TASKS

Task Number	Task Title
071-030-0004	Engage Targets with an MK 19 Grenade Machine Gun
071-054-0004	Engage Targets with an M136 Launcher
071-325-4407	Employ Hand Grenades
071-311-2130	Engage Targets with an M203 Grenade Launcher
071-010-0006	Engage Targets with an M249 Machine Gun
071-025-0007	Engage Targets with an M240B Machine Gun
071-100-0030	Engage Targets with an M16-Series Rifle/M4-Series Carbine
071-326-0502	Move Under Direct Fire
071-326-5611	Conduct the Maneuver of a Squad
071-326-5630	Conduct Movement Techniques by a Platoon
071-121-4080	Send a Spot Report (SPOTREP)
061-283-1011	Engage Targets with Indirect Fires
113-571-1022	Perform Voice Communications

SUPPORTED COLLECTIVE TASKS

Task Number	Task Title
07-2-1090	Conduct a Movement to Contact (Platoon-Company)
07-2-1450	Secure Routes (Platoon-Company)
07-2-9002	Conduct a Bypass (Platoon-Company)
07-2-9009	Conduct a Withdrawal (Platoon-Company)

TASK: React to an Obstacle (17-3-D8008)

CONDITIONS: The platoon is conducting tactical operations as part of a higher unit and has communication with the commander. The platoon or a section/squad makes contact with an obstacle. The platoon may or may not have countermine equipment. Enemy contact is possible. Some iterations of this task should be conducted in MOPP4 and under conditions of limited visibility.

CUE: Any Soldier gives an oral or visual signal they are in contact with an obstacle.

STANDARDS: The platoon identifies the obstacle, deploys as applicable to avoid decisive engagement of the entire platoon, and alerts the higher unit of obstacle contact and location. Once the obstacle is breached or bypassed, the platoon remains prepared to continue the unit mission. No friendly unit suffers casualties or equipment damage as a result of fratricide.

TASK STEPS AND PERFORMANCE MEASURES:

1. If applicable, element in contact with the obstacle alerts the platoon with a contact report.
2. In close direct fire contact situations, platoon takes immediate protective actions.
   a. Platoon leader (PL) directs the platoon to deploy to a covered and concealed location.
   b. As applicable, element in contact employs onboard smoke grenades and direct fire to obscure and suppress the enemy forces overwatching the obstacle.
3. In out-of-contact situations (platoon identifies obstacle from a position of advantage), platoon takes immediate protective actions.
   a. PL directs the platoon to deploy to a covered and concealed location.
   b. Element in visual contact with obstacle establishes an overwatch position.
   c. As applicable, employs direct fire and/or indirect fire to obscure and suppress the enemy forces overwatching the obstacle.
4. PL/platoon sergeant takes actions to develop the situation and report to the commander.
   a. Sends contact report to the higher commander.
   b. Develops the situation by section/squad (maneuver) to determine location, composition, and disposition of enemy forces overwatching the obstacle.
      (1) Directs one section/squad to establish a suitable overwatch position to allow platoon to continue to develop the situation.

TASK STEPS AND PERFORMANCE MEASURES:

(2) Directs the other section/squad to perform reconnaissance of the obstacle to determine composition of the obstacle and to locate a bypass. **NOTE:** Reconnaissance may be performed mounted or dismounted.

c. Sends obstacle report to the higher commander describing type, width, length, effect, and location of the obstacle.

d. Sends updated situation reports to the higher commander as necessary.

5. If a bypass is possible, PL reports the location of the bypass to the higher commander and recommends bypassing the obstacle.

NOTE: Once ordered to bypass, the platoon executes steps to bypass the obstacle. (Refer to task 07-2-9002, Conduct a Bypass) [Platoon–Company]).

6. If a bypass is not possible, PL reports to the higher commander and recommends, based on obstacle composition, a point of breach and either platoon-level reduction or a higher-level breach.

NOTE: If ordered to reduce the obstacle, the platoon executes steps of breach force operations. (Refer to task 17-2-3070, Breach an Obstacle [Platoon–Company]).

SUPPORTING PRODUCTS

Product ID	Product Name
FM 3-20.15	Tank Platoon
FM 3-90.1	Armor and Rifle Company Team

SUPPORTING INDIVIDUAL TASKS

Task Number	Task Title
171-121-3009	Control Techniques of Movement
171-121-4009	Conduct Scout Platoon Actions on Contact
171-121-4010	Conduct Tank Platoon Actions on Contact
171-121-4017	Supervise Tank Platoon Formations and Drills
171-121-4038	Supervise Local Security
171-121-4045	Conduct Troop Leading Procedures
171-121-4059	Conduct an Armor in-Stride Breach of a Minefield
171-121-4068	Conduct a Reconnaissance by Fire
071-100-0030	Engage Targets with an M16-Series Rifle/M4-Series Carbine
071-325-4407	Employ Hand Grenades
071-326-0501	Move as a Member of a Fire Team
071-326-0608	Use Visual Signaling Techniques

SUPPORTING INDIVIDUAL TASKS

Task Number   **Task Title**

071-326-0503   Move Over, Through, or Around Obstacles (Except Minefields)

SUPPORTED COLLECTIVE TASK

Task Number   **Task Title**

07-2-1090   Conduct a Movement to Contact (Platoon-Company)

17-2-3070   Breach an Obstacle (Platoon-Company)

17-2-4000   Conduct Route Reconnaissance (Platoon-Company)

TASK: Establish Security at the Halt (07-3-D9508)

CONDITIONS:

Dismounted/mounted. The unit moves tactically, conducting operations. An unforeseen event causes the unit to halt. Enemy contact is possible.

CUE: This drill begins when the unit must halt and enemy contact is possible, or the unit leader initiates drill by giving the order, "HALT!"

STANDARDS:

Dismounted. Soldiers stop movement and clear the area per unit standing operating procedures (SOPs). (An example technique is the 5-25 meters; each Soldier immediately scans 5 meters around his position and then searches out to 25 meters based on the duration of the halt). Soldiers occupy covered and concealed positions, and maintain dispersion and all-round security.

Mounted. Vehicle commanders direct their vehicles into designated positions, using available cover and concealment. Soldiers dismount in the order specified and clear the area per unit SOPs. (An example technique is the 5-25 meters; each Soldier immediately scans 5 meters around his position and then searches out to 25 meters based on the duration of the halt). Platoon/section members maintain dispersion and all-round security.

TASK STEPS AND PERFORMANCE MEASURES

1. **Dismounted.** Unit personnel take the following actions:

    a. The unit leader gives the arm-and-hand signal to halt.

    b. Soldiers establish local security. They take the following actions:

    (1) Assume hasty fighting positions using available cover and concealment.

    (2) Inspect and clear their immediate area (Example: using the 5-25 technique).

    (3) Establish a sector of fire for their assigned weapon (Example: using 12 o'clock as the direction the Soldier is facing, the Soldier's sector of fire could be his 10 o'clock to 2 o'clock).

    c. Element leaders adjust positions as necessary. They take the following actions:

    (1) Inspect and clear their element area.

    (2) Ensure Soldiers sector of fire overlap.

    (3) Coordinate sectors with the elements on their left and right.

    d. Unit leaders report the situation to higher headquarters (HQ).

2. **Mounted.** Unit personnel take the following actions:

    a. Unit leaders give the order over the radio to stop movement and establish security (See Figure 1.)

TASK STEPS AND PERFORMANCE MEASURES

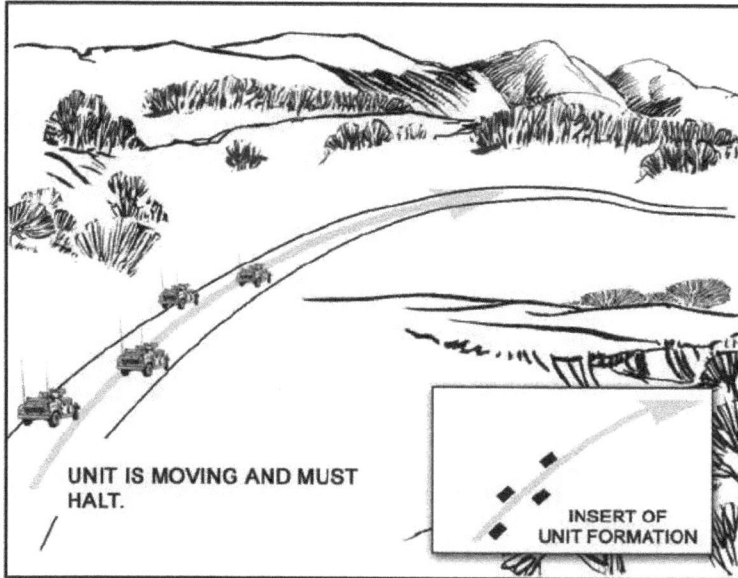

UNIT IS MOVING AND MUST HALT.

INSERT OF
UNIT FORMATION

Figure 1. Establish security at the halt

b.   The unit halts in the herringbone or coil formation according to the unit SOPs. (See Figures 2 and 3.)

TASK STEPS AND PERFORMANCE MEASURES

Figure 2. Establish security at the halt (mounted) (herringbone) (continued)

TASK STEPS AND PERFORMANCE MEASURES

VEHICLES MOVE INTO POSITIONS AND
SOLDIERS DISMOUNT TO PROVIDE
SECURITY.

Figure 3. Establish security at the halt (mounted) (coil) (continued)

c. Each vehicle commander ensures his vehicle is correctly positioned, using cover and concealment, and the crew served weapon is manned and scanning.

d. Vehicle commanders order Soldiers to dismount to provide local security.

e Soldiers dismount and establish local security. They take the following actions:

(1) Move to a covered and concealed position as designated by the leader.

(2) Inspect and clear their immediate area (example: using the 5-25 technique).

(3) Establish a sector of fire for their assigned weapons.

f. Dismount element leaders adjust positions as necessary.

g. Unit leaders report the situation to higher HQ.

SUPPORTING PRODUCTS

Product ID	Product Name
FM 3-21.8	The Infantry Rifle Platoon and Squad
FM 3-21.75	Warrior Ethos and Soldier Combat Skills
ATTP 3-21.9	SBCT Infantry Rifle Platoon and Squad

SUPPORTING INDIVIDUAL TASKS

Task Number	Task Title
113-571-1022	Perform Voice Communications
551-88M-0005	Operate a Vehicle in a Convoy
071-326-0513	Select Temporary Fighting Positions
071-326-0608	Use Visual Signaling Techniques
071-331-0801	Challenge Persons Entering Your Area
071-331-0815	Practice Noise, Light, and Litter Discipline
071-331-1004	Perform Duty as a Guard
191-376-4114	Control Entry to and Exit From a Restricted Area
191-376-5140	Search a Vehicle for Explosive Devices or Prohibited Items at an Installation Access Control Point
191-376-5151	Control Access to a Military Installation
551-001-1040	Perform 5/25-Meter Scans
551-001-1041	Establish Security While Mounted (If Applicable)
551-001-1042	Dismount a Vehicle
551-001-1043	React to Vehicle Rollover
551-88M-1658	Prepare Vehicle for Convoy Operations

SUPPORTED COLLECTIVE TASKS

Task Number	Task Title
19-3-2007	Conduct Convoy Security
07-2-1189	Conduct a Dismounted Tactical Road March (Platoon-Company)
07-2-1198	Conduct a Mounted Tactical Road March (Platoon-Company)
07-2-1342	Conduct Tactical Movement (Platoon-Company)
07-2-9011	Conduct Tactical Movement in an Urban Area (Platoon-Company)

TASK: React to an IED Attack While Maintaining Movement (05-3-D0017)

CONDITIONS: The element conducts a mounted military operation when an improvised explosive device (IED) detonates.

CUE: An IED detonates within casualty-producing radius on the patrol, resulting in varying degrees of battle damage to the vehicles, equipment, and personnel.

STANDARDS: React to the IED attack by performing 5/25 meter checks. They use the 5 Cs (confirm, clear, cordon, check, and control) to suppress enemy fire, set up security, evacuate casualties, recover disabled vehicles, submit an explosive hazards spot report, and exit the area.

TASK STEPS AND PERFORMANCE MEASURES

Unit personnel take the following actions:

1. Report the IED attack to the patrol (any Soldier can do this using the 3 Ds: distance, direction, and description).
2. Establish 360-degree local security by directing the element to focus outward from the attack site.
3. If necessary, direct the element to the rally point based upon mission, enemy, terrain and weather, troops and support available, time available, and civil considerations (METT-TC) factors.
4. Employ tactical combat casualty care measures.
5. Evacuate casualties.
6. Conduct consolidation and reorganization at the rally point.
7. Direct element members to report the status of liquid, ammunition, casualties, and equipment (LACE) report.

SUPPORTING PRODUCTS

Product ID	Product Name
FM 3-21.75	Warrior Ethos and Soldier Combat Skills
ATTP 3-21.9	SBCT Infantry Rifle Platoon and Squad

SUPPORTING INDIVIDUAL TASKS

Task Number	Task Title
071-030-0004	Engage Targets with an MK 19 Grenade Machine Gun
071-054-0004	Engage Targets with an M136 Launcher
071-325-4407	Employ Hand Grenades
071-311-2130	Engage Targets with an M203 Grenade Launcher
071-010-0006	Engage Targets with an M249 Machine Gun
071-025-0007	Engage Targets with an M240B Machine Gun

SUPPORTING INDIVIDUAL TASKS

Task Number	Task Title
071-100-0030	Engage Targets with an M16-Series Rifle/M4-Series Carbine
071-326-0502	Move Under Direct Fire
071-326-5611	Conduct the Maneuver of a Squad
071-326-5630	Conduct Movement Techniques by a Platoon
071-121-4080	Send a Spot Report (SPOTREP)
061-283-1011	Engage Targets with Indirect Fires
113-571-1022	Perform Voice Communications

SUPPORTED COLLECTIVE TASKS

Task Number	Task Title
07-2-1090	Conduct a Movement to Contact (Platoon-Company)
07-2-1450	Secure Routes (Platoon-Company)
07-2-9002	Conduct a Bypass (Platoon-Company)
07-2-9009	Conduct a Withdrawal (Platoon-Company)

TASK: Establish a Hasty Checkpoint (19-4-D0105)

CONDITIONS The element receives an order from higher headquarters (HQ) to immediately establish a hasty checkpoint (CP) at a specific location in its area of operations (AO). The local police or security forces may assist with the operations. The unit receives guidance on the rules of engagement (ROE), rules of interaction (ROI), and escalation of force (EOF). Translators or host-nation personnel are attached or available. Some iterations of this task should be performed in mission-oriented protective posture 4 (MOPP4).

CUE: This drill begins when the element leader receives the order from higher HQ and issues the command to the element to immediately establish a hasty checkpoint (CP) at a specified location, or the element leader directs his personnel to conduct this drill.

STANDARDS: The element takes immediate action to construct and establish a hasty CP according to orders from higher HQ. The checkpoint controls vehicular and pedestrian traffic by limiting entry to and exit from the specified area. The element is briefed on the ROE, ROI, and EOF, mission instructions, higher HQ order, and other special orders. The time required to perform this task is increased when conducting it in MOPP4.

TASK STEPS AND PERFORMANCE MEASURES

1.   The element leader receives and issues orders to element to immediately establish a hasty CP at a specific location. He takes the following actions:

NOTE: Establish a hasty CP when the CP will be used for a set period of time, usually a short duration. Hasty CPs should be located as to achieve the element of surprise and cannot be seen by approaching traffic until it is too late to withdraw. Good locations for hasty CPs are bridges, defiles, highway intersections, reverse slopes of hills, and just beyond sharp curves.

   a. Issues directives and assignments to personnel to immediately set up and construct key elements of a hasty CP. The key elements are:

   (1) Establish security.
   (2) Establish security positions or occupy sentry positions.
   (3) Establish communication.
   (4) Construct an entry point.
   (5) Construct approach lanes and protective barriers.
   (6) Construct search and holding areas.
   (7) Establish lethal and non-lethal (weapons) overwatch positions.
   (8) Post warning signs.

TASK STEPS AND PERFORMANCE MEASURES

(9) Final operations briefing and instructions (prior to activating the CP).

b. Issues time-line for establishing the checkpoint.

2. Element members immediately perform their assigned duties to establish or construct the hasty CP to prepare it for operation (based on the element leader's instructions). They take the following actions:

NOTE: Security, hasty defensive positions, and establishing commo must be completed first. The remaining hasty CP construction duties and responsibilities are not required to be performed in sequence and can be completed simultaneously by various element members to speed up the process.

a. Establish security during the construction of the hasty checkpoint.

b. Construct hasty defensive positions (all members).

c. Establish communications with all elements and higher HQ.

d. Establish an initial (stand-off) visual search area or zone (only if mission dictates or allows) that is clearly marked with signs for vehicles and/or pedestrians to stop and wait for further instructions prior to approaching the actual CP entrance.

NOTE: The initial search zone is a distant visual search area where vehicles and personnel will be ordered (by visual or audio means) to stop at a clearly marked point before they actually enter the CP. This process may allow for the detection of weapons and explosives at a safe distance or cause a person to reveal their intent. Personnel and vehicles can be visually inspected from a predetermined distance (approximately 25 to 100 meters or as the mission dictates) while CP operators remain behind a protective barrier or vehicle. Personnel are ordered to exit their vehicle, open their vehicle compartments that can be observed from a distance (trunk, hood, etc), uncover or take out items from their vehicle, open or pull up their overgarments, turn around, and perform any other additional measures according to the SOP. This visual search is conducted prior to bringing personnel and vehicles into the checkpoint for a detailed search. Local support authorities can be utilized for this area. The initial search zone is more applicable to a deliberate CP but can be used for a hasty CP if the mission allows and threat dictates its use.

e. Construct entry points that can restrict and control the entry of vehicles and/or pedestrians into the hasty CP. Entry point should also provide minimal protection for CP personnel if needed.

TASK STEPS AND PERFORMANCE MEASURES

NOTE: Hasty CP approach lanes, entry point, protective barriers, holding areas, and search areas can be constructed utilizing readily available materials such as engineer tape, debris, trees, rocks, concertina wire, existing structures, and all other available equipment/material. The element should also use existing culverts, bridges, deep cuts, sharp bends, or dips in the road to create the hasty checkpoint. Ensure that there is adequate lighting for night operations (if applicable).

f. Create or construct approach (canalization) and deceleration lanes that force traffic to slow down, and directs vehicles and/or pedestrians to the designated areas. This can be done with a system of curves and obstacles vehicles must maneuver around as they approach the CP.

g. Establish or construct holding areas for detained persons away from checkpoint entrance.

NOTE: The holding and search areas are relatively secure areas where personnel and vehicles are positively identified and a complete detailed search is conducted. Existing structures, vehicles or obstacles are used to isolate vehicles or individuals from others with overwatch protection from weapon positions. Mission may require male and female members to conduct personnel searches (female soldiers should conduct searches of female personnel entering the CP).

h. Establish or construct detailed search areas for personnel (male and female) and vehicles.

i. Establish lethal and nonlethal overwatch positions where they can observe all areas and approach areas. Crew served weapons will be strategically placed at these locations.

j. Ensure warning signs are posted.

NOTE: Warning or instructional signs should be posted in the native and English languages in the CP area. Signs should be placed at key locations and distances leading up to CP. Signs should also specify when deadly force is authorized for failure to comply with posted warnings (based on the current SOPs, orders, ROE, and EOF).

3. The element leader conducts final briefing, after hasty CP is established, and makes notification to HQ. He takes the following actions:

a. Briefs personnel on key elements of conducting hasty CP operations (prior to putting the CP into operation). The key elements are:

(1) Current ROE, EOF, and ROI.

(2) Rules regarding search, detention, standoff distances, and the use of force.

(3) Actions on contact (mounted and dismounted threats).

(4) Procedures for clearing and processing personnel and vehicles through the CP according to orders and guidance from higher HQ.

TASK STEPS AND PERFORMANCE MEASURES

(5) Utilizing all available assets and procedures the mission allows to conduct CP operations safely (stand-off distance, obstacles, barriers, and warning signs).

(6) Searching for high risk and prohibited items (such as weapons, explosives, and contraband).

(7) Maintaining security and overwatch support of CP.

(8) Ensuring vehicle traffic, movement, and personnel are handled according to current directives and SOPs.

b. Notifies higher HQ that the hasty CP is established and ready for operations.

SUPPORTING PRODUCTS

Product ID	Product Name
FM 3-21.75	Warrior Ethos and Soldier Combat Skills

SUPPORTED INDIVIDUAL TASKS

Task Number	Task Title
081-831-0101	Request Medical Evacuation
081-831-1003	Perform First Aid to Clear an Object Stuck in the Throat of a Conscious Casualty
081-831-1005	Perform First Aid to Prevent or Control Shock
081-831-1007	Perform First Aid for Burns
081-831-1025	Perform First Aid for an Open Abdominal Wound
081-831-1026	Perform First Aid for an Open Chest Wound
081-831-1032	Perform First Aid for Bleeding and/or Severed Extremity
081-831-1033	Perform First Aid for an Open Head Wound
081-831-1034	Perform First Aid for a Suspected Fracture
081-831-1046	Transport a Casualty
113-571-1022	Perform Voice Communications
805C-PAD-2060	Report Casualties

SUPPORTED COLLECTIVE TASKS

Task Number	Task Title
07-2-5027	Conduct Consolidation and Reorganization (Platoon-Company)
08-2-0003	Treat Casualties
08-2-0004	Evacuate Casualties

TASK: React to Indirect Fire (07-3-D9504)

CONDITIONS:

Dismounted. The unit moves, conducting operations. Any Soldier gives the alert, INCOMING, or a round impacts nearby.

Mounted. The platoon/section is stationary or moves, conducting operations. The alert, INCOMING, comes over the radio or intercom or rounds impact nearby.

CUE: This drill begins when any member alerts, INCOMING, or a round impacts.

STANDARDS:

Dismounted. Soldiers immediately seek the best available cover. The unit moves out of area to the designated rally point after the impacts.

Mounted. When moving, drivers immediately move their vehicles out of the impact area in the direction and distance ordered. If stationary, drivers start their vehicles and move in the direction and distance ordered. Unit leaders report the contact to higher headquarters.

TASK STEPS AND PERFORMANCE MEASURES

1. **Dismounted.** Unit personnel take the following actions:
   a. Any Soldier announces, "INCOMING!"
   b. Soldiers immediately assume the prone position or move to immediate available cover during initial impacts.
   c. The unit leader orders the unit to move to a rally point by giving a direction and distance.
   d. After the impacts, Soldiers move rapidly in the direction and distance to the designated rally point.
   e. The unit leaders report the contact to higher HQ.
2. **Mounted.** Unit personnel take the following actions:
   a. Any Soldier announces, "INCOMING!"
   b. Vehicle commanders repeat the alert over the radio.
   c. The leaders give the direction and link-up location over the radio.
   d. Soldiers close all hatches if applicable to the vehicle type; gunners stay below turret shields or get down into vehicle.
   e. Drivers move rapidly out of the impact area in the direction ordered by the leader.
   f. Unit leaders report the contact to higher HQ.

SUPPORTING PRODUCTS

Product ID	Product Name
FM 3-21.8	The Infantry Rifle Platoon and Squad

SUPPORTING PRODUCTS

Product ID   **Product Name**
FM 3-21.75   Warrior Ethos and Soldier Combat Skills
ATTP 3-21.9   SBCT Infantry Rifle Platoon and Squad

SUPPORTING INDIVIDUAL TASKS

Task Number   **Task Title**
071-326-3002   React to Indirect Fire While Mounted
113-571-1022   Perform Voice Communications

SUPPORTED COLLECTIVE TASKS

Task Number   **Task Title**
07-2-3000   Conduct Support by Fire (Platoon-Company)
07-2-9004   Conduct a Delay (Platoon-Company)
07-2-9009   Conduct a Withdrawal (Platoon-Company)
17-2-9225   Conduct a Screen (Platoon-Company)

TASK: Conduct the 5 Cs (05-3-D0016)

CONDITIONS: The element conducts a mounted or dismounted military patrol when an improvised explosive device (IED) is identified or detonates.

CUE: This is done when a possible or suspected IED is identified, an explosive device is detonated, or while conducting a security halt (mounted or dismounted).

STANDARDS: The element conducts the 5 Cs (confirm, clear, cordon, check, control) correctly, ensuring the area is clear of any nonessential personnel, secondary or tertiary IEDs have been confirmed and identified, a cordon has been established, and personnel access to the area is under control.

TASK STEPS AND PERFORMANCE MEASURES

NOTE: Conduct the 5 Cs; these are not order specific and can be done concurrently.

1. Confirms there is a requirement for explosive ordnance disposal (EOD) when encountering a suspected or known IED.
2. Clears all personnel from the area to a tactically safe position and distance from the potential IED.
3. Cordons the area.
4. Checks the immediate area for secondary/tertiary devices around the incident control point (ICP) and cordon using the 5/25 meter checks.
5. Controls the area inside the cordon to ensure only authorized access.

> **DANGER**
> MINIMUM SAFE DISTANCE FOR EXPOSED PERSONNEL IN THE OPEN IS 300 METERS.

SUPPORTING PRODUCTS

Product ID	Product Name
FM 3-21.75	Warrior Ethos and Soldier Combat Skills

SUPPORTED INDIVIDUAL TASKS

Task Number	Task Title
081-831-0101	Request Medical Evacuation
081-831-1003	Perform First Aid to Clear an Object Stuck in the Throat of a Conscious Casualty

SUPPORTED INDIVIDUAL TASKS

Task Number	Task Title
081-831-1005	Perform First Aid to Prevent or Control Shock
081-831-1007	Perform First Aid for Burns
081-831-1025	Perform First Aid for an Open Abdominal Wound
081-831-1026	Perform First Aid for an Open Chest Wound
081-831-1032	Perform First Aid for Bleeding and/or Severed Extremity
081-831-1033	Perform First Aid for an Open Head Wound
081-831-1034	Perform First Aid for a Suspected Fracture
081-831-1046	Transport a Casualty
113-571-1022	Perform Voice Communications
805C-PAD-2060	Report Casualties

SUPPORTED COLLECTIVE TASKS

Task Number	Task Title
07-2-5027	Conduct Consolidation and Reorganization (Platoon-Company)
08-2-0003	Treat Casualties
08-2-0004	Evacuate Casualties

TASK: Evacuate a Casualty (Dismounted and Mounted) (07-3-D9507)

CONDITIONS: The unit is stationary or moves, conducting operations. A Soldier has been injured and must be evacuated. All enemy in the area have been suppressed, neutralized, or destroyed, and local security is established. Some iterations of this drill should be performed in mission-oriented protective posture 4 (MOPP 4).

CUE: This drill begins when a unit member is injured and must be evacuated or the leader directs his personnel to conduct the drill.

STANDARD: Element members conduct first aid and evacuate the casualties without dropping or causing further injury to the casualties. If necessary, the unit leader, combat medic, or any Soldier requests medical evacuation (MEDEVAC) and reports the contact to higher headquarters (HQ).

TASK STEPS AND PERFORMANCE MEASURES

1. Element members conduct first aid and evacuate the casualties without dropping or causing further injury to the casualties.
2. Drill is conducted while dismounted. Unit personnel take the following actions:
    a. Any unit member provides initial first aid (self-aid/buddy aid).
    b. Any unit combat lifesaver provides enhanced first aid or combat medic provides emergency medical treatment, if necessary.
    c. The unit leader, combat medic, or any Soldier requests MEDEVAC using the 9-Line MEDEVAC request, if necessary.
    d. The unit aid and litter team or designated members evacuate casualties to the casualty collection point (CCP) or patient collecting point (PCP) and request MEDEVAC. They take the following actions:
        (1) Remove key operational items and equipment (maps, simple key loader [SKL]/automated network control devices [ANCD], position-locating devices, laser pointers, and other sensitive items).
        (2) Account for the weapons and ammunition of casualties according to the unit standing operation procedures (SOPs).
        (3) Complete DD Form 1380, *U.S. Field Medical Card*, and unit leaders or any member complete Department of the Army (DA) Form 1156, *Casualty Feeder Card.*
        (4) Evacuate casualty to the CCP, PCP, or aid station using litters, one or two man carry, or by having casualties with minor wounds walk.
3. Drill is conducted while mounted. Unit personnel take the following actions:
    a. Crew/occupants provide initial first aid (self-aid/buddy aid).

TASK STEPS AND PERFORMANCE MEASURES

b. Any unit combat lifesaver, combat medic, or designated Soldier moves to the vehicle to provide first aid or enhanced first aid (self-aid, buddy aid, and combat lifesaver) and emergency medical treatment (EMT) (combat medic) and then evacuates the casualty.

c. Designated Soldiers remove the casualty from the vehicle so as not to cause further injury. They take the following actions:

(1) Remove all key operational items and equipment (maps, simple key loader [SKL]/automated network control devices [ANCD], position-locating devices, and all other sensitive items).

(2) Account for the weapons and ammunition of casualties according to unit SOPs.

(3) Complete DD Form 1380 and DA Form 1156.

(4) Evacuate casualties to the CCP or PCP and request MEDEVAC (9-line MEDEVAC request) or evacuate directly to the aid station using available vehicle assets.

4. Unit leaders report the contact and casualties according to unit SOPs to higher HQ.

SUPPORTING PRODUCTS

Product ID	Product Name
FM 3-21.8	The Infantry Rifle Platoon and Squad
FM 3-21.75	Warrior Ethos and Soldier Combat Skills

SUPPORTED INDIVIDUAL TASKS

Task Number	Task Title
081-831-0101	Request Medical Evacuation
081-831-1003	Perform First Aid to Clear an Object Stuck in the Throat of a Conscious Casualty
081-831-1005	Perform First Aid to Prevent or Control Shock
081-831-1007	Perform First Aid for Burns
081-831-1025	Perform First Aid for an Open Abdominal Wound
081-831-1026	Perform First Aid for an Open Chest Wound
081-831-1032	Perform First Aid for Bleeding and/or Severed Extremity
081-831-1033	Perform First Aid for an Open Head Wound
081-831-1034	Perform First Aid for a Suspected Fracture
081-831-1046	Transport a Casualty
113-571-1022	Perform Voice Communications
805C-PAD-2060	Report Casualties

SUPPORTED COLLECTIVE TASKS

Task Number	Task Title
07-2-5027	Conduct Consolidation and Reorganization (Platoon-Company)
08-2-0003	Treat Casualties
08-2-0004	Evacuate Casualties

TASK: Enter a Trench to Secure a Foothold (07-3-D9410)

CONDITIONS: The platoon moves tactically and receives effective fire from an enemy trench. The platoon is ordered to secure a foothold in the trench. The platoon has only organic weapons support available.

CUE: The platoon leader initiates drill by giving the order for the assault element to secure a foothold in the trench.

STANDARDS: The platoon leader quickly identifies the entry point. The platoon secures the entry point, enters the trench, and secures an area large enough for the follow-on unit. The platoon maintains a sufficient fighting force to repel enemy counterattack and continues the mission.

TASK STEPS AND PERFORMANCE MEASURES

1.   A platoon executes actions on contact to eliminate or suppress fires from the trench.
2.   The section/squad in contact takes the following actions:
    a. Deploys; takes the following actions:
      (1)   Returns fire.
      (2)   Seeks cover.
      (3)   Establishes fire superiority.
      (4)   Establishes local security.
      (5)   The platoon sergeant repositions other sections/squads to focus supporting fires and increase observation.
    b. Reports; takes the following actions:
      (1)   Section/squad leader reports location of hostile fire to platoon leader from base-of-fire position using the size, activity, location, unit, time, and equipment (SALUTE) format.
      (2)   The platoon leader sends contact report followed by a SALUTE report to commander.
3.   The platoon leader evaluates and develops the situation. He takes the following actions:
    a. Evaluates the situation using the situation reports (SITREPs) from the section/squad in contact and his personal observations. At the minimum his evaluation should include—
      (1)   Number of enemy weapons or volume of fire.
      (2)   Presence of vehicles.
      (3)   Employment of indirect fires.
    b. The platoon leader quickly develops the situation by taking the following actions:
      (1)   Conducts a quick reconnaissance to determine enemy flanks.
      (2)   Locates mutually supporting positions.

TASK STEPS AND PERFORMANCE MEASURES

(3) Locates any obstacles that impede the assault or provide some type of cover or concealment.

(4) Determines whether the force is inferior or superior.

(5) Analyzes reports from section/squad leaders, teams in contact, or adjacent units.

4. The platoon leader chooses a course of action (COA). He takes the following actions:

a. Decides to enter the trench and selects his entry point.

b. Selects a covered and concealed route to his entry point.

c. Directs his maneuver element to secure the near side of the entry point and reduce the obstacle to gain a foothold.

d. Repositions the remaining section/squad to provide additional observation and supporting fires.

5. The platoon executes COA (uses suppress, obscure, secure, reduce, assault [SOSRA] to set conditions for the assault). (See Figure 1.) It takes the following actions:

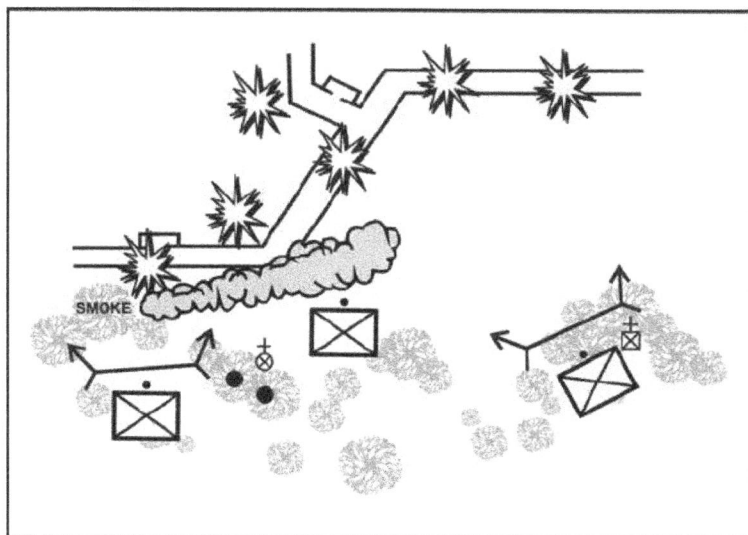

Figure 1. Enter a trench to secure a foothold

a. Suppresses and obscures. Takes the following actions:

(1) Ensures platoon leader or forward observer (FO) calls for and adjusts indirect fire in support of assault.

(2) Ensures platoon sergeant directs base-of-fire section/squad to cover maneuvering section/squad.

TASK STEPS AND PERFORMANCE MEASURES

(3) Obscures maneuver element's movement with smoke, if available.

b. Secures the near side and reduce the obstacle. The maneuver section/squad clears entry point. They take the following actions:

(1) Section/squad leader moves the assaulting squad to last covered and concealed position short of the entry point.

(2) Section/squad leader designates entry point.

(3) Base-of-fire section/squad shifts fires from entry point and continues to suppress adjacent enemy positions.

(4) Section/squad leader uses one team to suppress the entry point and positions the assaulting team at the entry point.

c. The platoon leader directs FO to shift indirect fires to isolate the object and the base of fire sections/squads to shifts fire as assault section/squad advances.

d. The platoon secures the far side and establishes a foothold. (See Figure 2.) It takes the following actions:

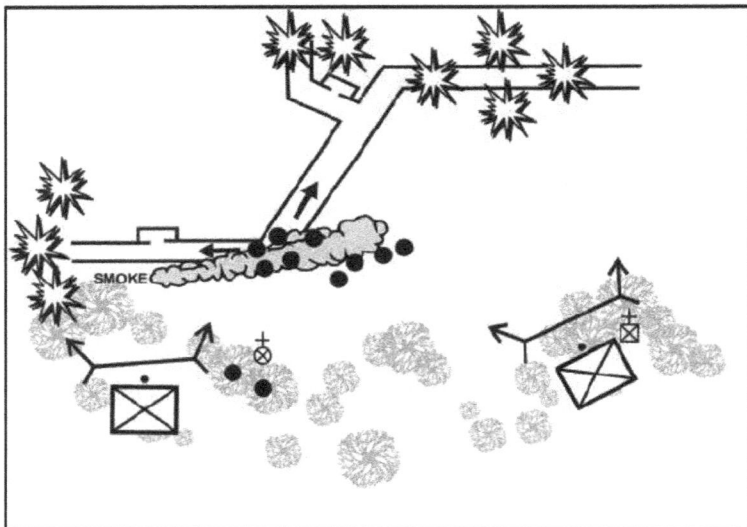

Figure 2. Enter a trench to secure a foothold (continued)

(1) Two Soldiers position themselves against the edge of the trench to roll right and left of the entry point to clear far side of obstacle and establish foothold.

TASK STEPS AND PERFORMANCE MEASURES

(2) The assault team engages all identified or likely enemy positions with rapid, short bursts of automatic fire and scanned the trench for concealed enemy positions. The rest of the section/squad provides immediate security outside the trench.

(3) The assault team clears enough room for the section/squad or to the first trench junction and announces, "CLEAR!"

(4) The section/squad leader marks entry point according to platoon standing operating procedures (SOPs), then sends next assault team in to increase the size of the foothold by announcing, "NEXT TEAM IN!"

(5) The next assault team moves into trench and secures assigned area. (See Figure 3.)

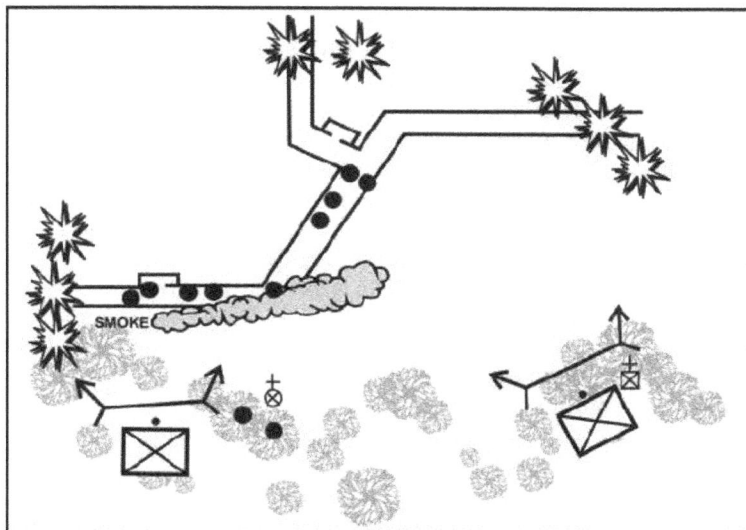

Figure 3. Enter a trench to secure a foothold (continued)

(6) The section/squad leader reports to platoon leader that the foothold is secure.

(7) The platoon leader moves to the maneuver section/squad leader to assess the situation.

(8) The platoon sergeant moves forward to control supporting squads outside the trench.

(9) The platoon leader sends necessary teams to clear an area large enough for the platoon, and then reports to the commander that the foothold is secure and if additional support is needed to continue clearing the trench.

6. The platoon/section/squad leaders account for Soldiers, provide a SITREP to higher HQ, reorganize as necessary, and continue the mission.

SUPPORTING PRODUCTS

Product ID	Product Name
FM 3-21.8	The Infantry Rifle Platoon and Squad
FM 3-21.75	Warrior Ethos and Soldier Combat Skills

SUPPORTING INDIVIDUAL TASKS

Task Number	Task Title
071-054-0004	Engage Targets with an M136 Launcher
071-325-4407	Employ Hand Grenades
071-326-0501	Move as a Member of a Fire Team
071-326-0502	Move Under Direct Fire
071-326-0503	Move Over, Through, or Around Obstacles (Except Minefields)
071-326-0513	Select Temporary Fighting Positions
071-326-0608	Use Visual Signaling Techniques
071-326-5605	Control Movement of a Fire Team
071-100-0030	Engage Targets with an M16-Series Rifle/M4-Series Carbine
071-025-0007	Engage Targets with an M240B Machine Gun
071-010-0006	Engage Targets with an M249 Machine Gun
071-410-0019	Control Organic Fires

SUPPORTED COLLECTIVE TASKS

Task Number	Task Title
07-2-1261	Conduct an Attack in an Urban Area (Platoon-Company)
07-2-9001	Conduct an Attack (Platoon-Company)

TASK: Breach of a Mined Wire Obstacle (07-3-D9412)

CONDITIONS: The platoon encounters a mine wire obstacle preventing the company's movement. The platoon's forward movement is stopped by a wire obstacle reinforced with mines that cannot be bypassed. The enemy engages the platoon from positions on the far side of the obstacle. This drill begins when the unit's lead element encounters a mine wire obstacle, and the unit leader orders an element to breach the obstacle.

CUE: This drill begins when the unit's lead element encounters a mine wire obstacle, and the unit leader orders an element to breach the obstacle.

STANDARDS: The platoon breaches the obstacle and moves all personnel and equipment quickly through the breach. The platoon moves the support element and follow-on forces through the breach and maintains a sufficient fighting force to secure the far side of the breach.

TASK STEPS AND PERFORMANCE MEASURES
1.   A platoon's section/squad executes actions on contact to reduce effective fires from the far side of the obstacle.
2.   The section/squad in contact takes the following actions:
     a. Deploys; takes the following actions:
        (1) Returns fire.
        (2) Seeks cover.
        (3) Establishes fire superiority.
        (4) Establishes local security.
        (5) Platoon sergeant repositions other squads to focus supporting fires and increase observation.
     b.   Reports; takes the following actions:
        (1) Squad leader reports location of hostile fire to platoon leader from base-of-fire position using the size, activity, location, unit, time, and equipment (SALUTE) format.
        (2) Platoon leader sends contact report followed by a SALUTE report to commander.
3.   The platoon leader evaluates and develops the situation. He takes the following actions:
     a. Quickly evaluates the situation with using the situation reports (SITREPs) from the squad in contact and his personal observations. At a minimum his evaluation should include—
        (1) Number of enemy weapons or volume of fire.
        (2) Presence of vehicles.
        (3) Employment of indirect fires.
     b.   Quickly develops the situation. He takes the following actions:

TASK STEPS AND PERFORMANCE MEASURES

(1) Conducts a quick reconnaissance to determine enemy flanks.

(2) Locates mutually supporting positions.

(3) Locates obstacles that impede the assault or provide some type of cover or concealment.

(4) Determines whether the force is inferior or superior.

(5) Analyzes reports from squad leaders, teams in contact, or adjacent units.

4.   The platoon leader directs the vehicles (if available) and the squad in contact to support the movement of another squad to the breach point. He takes the following actions:

a. Indicates the route to the base-of-fire position.

b. Indicates the enemy position to be suppressed.

c. Indicates the breach point and the route the rest of the platoon will take.

d. Gives instructions for lifting and shifting fires.

5.   On the platoon leader's signal, the base-of-fire squad takes the following actions:

a. Destroys or suppresses enemy weapons that are firing effectively against the platoon.

b. Obscures the enemy position with smoke.

c. Continues to maintain fire superiority while conserving ammunition and minimizing forces in contact.

6.   The platoon leader designates one squad as the breach squad and the remaining squad as the assault squad once the breach has been made. (The assault squad may add its fires to the base-of-fire squad. Normally, it follows the covered and concealed route of the breach squad and assaults through immediately after the breach is made.)

7.   The base-of-fire squad moves to the breach point and establishes a base of fire.

8.   The platoon sergeant moves forward to the base-of-fire squad with the second machine gun and assumes control of the squad.

9.   The platoon leader leads the breach and assault squads along the covered and concealed route.

10. The platoon forward observer (FO) calls for and adjusts indirect fires as directed by the platoon leader to support the breach and assault.

11. The breach squad executes actions to breach the obstacle (footpath). The squad leader takes the following actions:

a. Directs one fire team to support the movement of the other fire team to the breach point.

b. Designates the breach point.

TASK STEPS AND PERFORMANCE MEASURES

c. Ensures the base-of-fire team continues to provide suppressive fires and to isolate the breach point. (See Figure 1.)

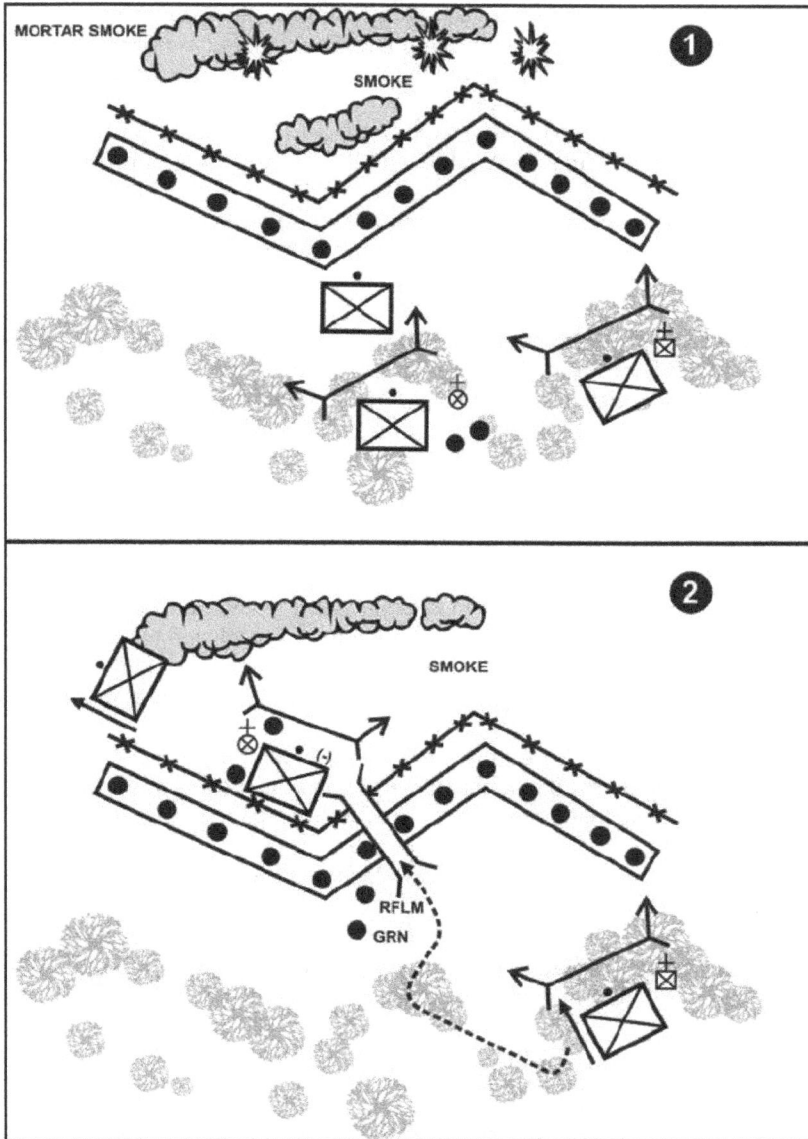

Figure 1. Breach obstacle

TASK STEPS AND PERFORMANCE MEASURES

d. The breaching fire team, with the squad leader, moves to the breach point using the covered and concealed route.

(1) The squad leader and breaching fire team leader employs smoke grenades to obscure the breach point. The platoon base-of-fire element shifts direct fires away from the breach point and continues to suppress adjacent enemy positions.

(2) The breaching fire team leader positions himself and the automatic rifleman on one flank of the breach point to provide close-in security.

(3) The grenadier and rifleman (or the antiarmor specialist and automatic rifleman) of the breaching fire team probe for mines and cut the wire obstacle, marking their path as they proceed. (Bangalore is preferred, if available.)

(4) Once the obstacle is breached, the breaching fire team leader and the automatic rifleman moves to the far side of the obstacle using covered and concealed positions. They signal the squad leader when they are in position and ready to support.

e. The squad leader signals the base-of-fire team leader to move his fire team up and through the breach. He then moves through the obstacle and joins the breaching fire team, leaving the grenadier (or antiarmor specialist) and rifleman of the supporting fire team on the near side of the breach to guide the rest of the platoon through.

f. Using the same covered and concealed route as the breaching fire team, the base-of-fire team moves through the breach and to a covered and concealed position on the far side.

12. The breach squad leader reports the situation to the platoon leader and posts guides at the breach point.

13. The platoon leader leads the assault squad through the breach in the obstacle and positions it on the far side to support the movement of the remainder of the platoon or to assault the enemy position covering the obstacle.

14. The breaching squad continues to widen the breach to allow vehicles to pass through.

15. The platoon leader provides a SITREP to the company commander and directs his breaching squad to move through the obstacle. The platoon leader appoints guides to guide the company through the breach point.

SUPPORTING PRODUCTS

Product ID	Product Name
FM 3-21.8	The Infantry Rifle Platoon and Squad
FM 3-21.75	Warrior Ethos and Soldier Combat Skills
ATTP 3-21.9	SBCT Infantry Rifle Platoon and Squad

SUPPORTING INDIVIDUAL TASKS

Task Number	Task Title
052-192-3060	Conduct a Breach a Minefield
052-193-1013	Neutralize Booby Traps
071-311-2129	Correct Malfunctions of an M203 Grenade Launcher
071-311-2130	Engage Targets with an M203 Grenade Launcher
071-326-0502	Move Under Direct Fire
071-326-5606	Select an Overwatch Position
071-326-0503	Move Over, Through, or Around Obstacles (Except Minefields)
071-326-5611	Conduct the Maneuver of a Squad

SUPPORTED COLLECTIVE TASKS

Task Number	Task Title
17-2-3070	Breach an Obstacle (Platoon-Company)

TASK: React to Air Attack Drill (17-3-D8004)

CONDITIONS: While operating in a tactical environment, the platoon or section identifies threat aircraft, requiring it to take either passive or active air defense measures. The platoon is digitally connected (if equipped) with higher headquarters (HQ) via Force XXI Battle Command Brigade and Below (FBCB2). Some iterations of this task should be performed in mission-oriented protective procedures 4 (MOPP4).

CUE: Any Soldier gives an oral or visual signal for a chemical attack or when a chemical alarm activates.

STANDARDS: The platoon or section executes appropriate air defense measures and prevents the aircraft from effectively engaging and/or observing the platoon/section. The platoon reports to higher HQ. No friendly unit suffers casualties or equipment damage as a result of fratricide.

TASK STEPS AND PERFORMANCE MEASURES
1.   The vehicle or individual who identifies threat aircraft alerts the platoon with a contact report containing these elements:
   a. Contact.
   b. Bandit(s).
   c. Cardinal direction (specify: north, south, east, or west).
2.   Platoon/section leaders analyze situation to determine whether the platoon is in the direct path of and/or is the target of the threat aircraft. They take the following actions:
   a. Order passive air defense measures when the platoon/section is not in the path of or target of the threat aircraft.
OR
   b. Order active air defense measures when the platoon is in the path of or target of the threat aircraft.
3.   Platoons or sections execute passive air defense measures as necessary. They take the following actions:
   a. On order of platoon/section leaders, move to covered and concealed positions, maintaining a minimum of 100 meters between vehicles and halts.
   b. Prepare to engage on order of platoon/section leader.
   c. Scan for follow-on aircraft.
NOTE: Higher HQ may order the platoon or section to continue movement.
4. Platoons execute active air defense measures as necessary. They take the following actions:
   a. If in the direct path of flight, move away from the path of flight as fast as possible, moving at a 45-degree angle toward the attacking aircraft.

TASK STEPS AND PERFORMANCE MEASURES

b. Maintain at least 100-meter intervals and avoid creating a linear target for the attacking aircraft.

c. Orient on the aim point designated by the platoon/section leader and engage the aircraft with a high volume of machine gun fire using the proper lead technique for the type of aircraft and direction of movement.

d. Move quickly to covered and concealed positions and halts.

e. Remain in covered and concealed positions, as required.

f. Scan for follow-on aircraft.

5. Platoon leaders/platoon sergeants (PSGs) report situation to higher HQ as necessary. They send—

a. Spot report (SPOTREP).

b. (D) Updated situation reports (SITREP), as necessary.

NOTE: Task steps and performance measures prefaced with a (D) may be performed digitally according to the order and/or unit SOPs. When preformatted messages do not exist or are not appropriate, free text messages may be substituted for FBCB2 messages identified in task steps.

SUPPORTING PRODUCTS

Product ID	Product Name
FM 3-20.98	Reconnaissance and Scout Platoon
FM 3-20.971	Reconnaissance and Cavalry Troop
FM 3-20.15	Tank Platoon

SUPPORTING INDIVIDUAL TASKS

Task Number	Task Title
171-121-4017	Supervise Tank Platoon Formations and Drills
171-121-4051	Prepare a Situation Report (SITREP)
171-121-4057	Perform Techniques of Movement

SUPPORTED COLLECTIVE TASKS

Task Number	Task Title
07-2-3000	Conduct Support by Fire (Platoon-Company)
07-2-9001	Conduct an Attack (Platoon-Company)
07-2-9003	Conduct a Defense (Platoon-Company)
07-2-9012	Conduct a Relief in Place (Platoon-Company)
17-5-5585	Engage Multiple Machine Gun Targets on a M1-Series Tank
17-5-5590	Conduct Main Gun Misfire Procedures on a M1-Series Tank

SUPPORTED COLLECTIVE TASKS

Task Number	Task Title
17-5-5622	Engage Targets with the Main Gun from a M1-Series Tank
17-5-8006	React to an Antitank Guided Missile (ATGM)

Appendix A

Weapons and Antiarmor Company Unit Task List

The UTL shown in Table A-1 identifies collective tasks that the unit is organized, manned, and equipped to conduct according to their TOE. This list assists the commander in developing his supporting collective task list and documenting which tasks to train that support the BN METL. The commander may accept risk and not train the entire UTL. The task numbers and task titles are listed under each of the six warfighting functions.

Table A-1. Example weapons and antiarmor company UTL

Task Number	Title
Mission Command	
07-2-5081	Conduct Troop-leading Procedures (Platoon-Company)
07-2-5135	Operate a Command Post (Platoon-Company)
55-2-4806	Prepare Equipment for Deployment
55-2-4828	Plan Unit Deployment Activities Upon Receipt of a Warning Order
Movement & Maneuver	
07-2-1090	Conduct a Movement to Contact (Platoon-Company)
07-2-9001	Conduct an Attack (Platoon-Company)
07-2-1256	Conduct an Attack by Fire (Platoon-Company)
07-2-1261	Conduct an Attack in an Urban Area (Platoon-Company)
07-2-1234	Conduct an Airborne Assault (Platoon-Company)
07-2-1495	Conduct an Air Assault (Platoon-Company)
07-2-9002	Conduct a Bypass (Platoon-Company)
07-2-9003	Conduct a Defense (Platoon-Company)
07-2-1378	Defend in an Urban Area (Platoon-Company)
07-2-9004	Conduct a Delay (Platoon-Company)
07-2-9009	Conduct a Withdrawal (Platoon-Company)

Table A-1. Example weapons and antiarmor company UTL (continued)

Task Number	Title
07-2-9012	Conduct a Relief in Place (Platoon-Company)
07-2-3000	Conduct Support by Fire (Platoon-Company)
07-2-1387	Employ a Reserve Force (Platoon-Company)
07-2-3027	Integrate Direct Fires (Platoon-Company)
17-2-0320	Conduct Infiltration (Platoon-Company)
07-2-3018	Employ Snipers (Platoon-Company)
07-2-1324	Conduct Area Security (Platoon-Company)
07-2-1396,	Employ Obstacles (Platoon-Company)
17-2-9225	Conduct a Screen (Platoon-Company)
17-2-4010,	Conduct Zone Reconnaissance (Platoon-Company)
17-2-4011	Conduct Area Reconnaissance (Platoon-Company)
07-2-9006	Conduct a Passage of Lines as the Passing Unit (Platoon-Company)
07-2-9007	Conduct a Passage of Lines as the Stationary Unit (Platoon-Company)
19-3-2406	Conduct Roadblock and Checkpoint
07-2-5027	Conduct Consolidation and Reorganization (Platoon-Company)
07-2-9014	Occupy an Assembly Area (Platoon-Company)
07-2-5009	Conduct a Rehearsal (Platoon-Company)
07-2-1405	Establish an Outpost (Platoon-Company)
44-3-3220	Perform Passive Air Defense Measures
44-3-3221	Perform Active Air Defense Measures
07-2-1189	Conduct a Dismounted Tactical Road March (Platoon-Company)
07-3-9016	Establish an Observation Post
07-2-6063	Maintain Operations Security (Platoon-Company)
07-2-1198	Conduct a Mounted Tactical Road March (Platoon-Company)
07-2-1369	Cross a Water Obstacle (Platoon-Company)
07-2-1252	Conduct an Antiarmor Ambush (Platoon-Company)
07-2-1342	Conduct Tactical Movement (Platoon-Company)

Table A-1. Example weapons and antiarmor company UTL (continued)

Task Number	Title
07-2-1450	Secure Routes (Platoon-Company)
17-2-3070	Breach an Obstacle (Platoon-Company)
07-2-5036	Conduct Coordination (Platoon-Company)
07-2-6045	Employ Camouflage, Concealment, and Deception Techniques (Platoon-Company)
07-2-9005	Conduct a Linkup (Platoon-Company)
07-2-9010	Conduct an Ambush (Platoon-Company)
07-3-9013	Conduct Action on Contact
07-2-9008	Conduct a Raid (Platoon-Company)
07-2-9011	Conduct Tactical Movement in an Urban Area (Platoon-Company)
07-2-9051	Conduct a Cordon and Search (Platoon-Company)
07-3-1072	Conduct a Disengagement
07-3-1333	Knock Out a Bunker (Platoon-Squad)
07-3-9017	Conduct Actions at Danger Areas
07-3-9018	Enter and Clear a Building (Section-Platoon)
07-3-9020	Establish a Patrol Base
07-3-9021	Clear a Trench Line
07-3-9022	Conduct a Security Patrol (Platoon-Squad)
07-3-9023	Conduct a Presence Patrol (Platoon-Squad)
17-2-4000	Conduct Route Reconnaissance (Platoon-Company)
19-3-2007	Conduct Convoy Security
03-2-9226	Cross a Chemically Contaminated Area
19-3-3107	Process Detainee(s) at Point of Capture (POC)
Intelligence	
34-2-0010	Report Tactical Information
34-3-0001	Monitor Platoon Operational Status
34-3-0003	Maintain Operations Security
Fires	
07-2-3036	Integrate Indirect Fire Support (Platoon-Company)
19-3-4004	Conduct Civil Disturbance Control
03-2-9223	React to the Initial Effects of a Nuclear Attack
03-2-9203	React to a Chemical or Biological (CB) Attack

Table A-1. Example weapons and antiarmor company UTL (continued)

Task Number	Title
Sustainment	
63-2-4546,	Conduct Logistics Package (LOGPAC) Support
08-2-0003,	Treat Casualties
08-2-0004,	Evacuate Casualties
Protection	
03-2-9224	Conduct Operational Decontamination
07-2-4054	Secure Civilians During Operations (Platoon-Company)
03-2-9201	Implement CBRN Protective Measures
07-2-5063	Conduct Composite Risk Management (Platoon-Company)

Appendix B
Weapons and Antiarmor Higher Headquarters' METL

Table B-1 depicts an example of the weapons and antiarmor company's higher headquarters METL for an Infantry battalion of an IBCT. For current information refer to DTMS.

Table B-1. Example of Infantry battalion, IBCT METL

IN BN IBCT	MET (ART)
	TG (T&EO)
	Supporting Collective Task (T&EO)
ART 7.1.2	Conduct an Attack
07-6-1072	TG: Conduct a Movement to Contact (Battalion - Brigade)
07-6-1081	Conduct a Passage of Lines as Passing Unit (Battalion - Brigade)
07-6-6082	Conduct Mobility, Countermobility, and/or Survivability (Battalion-Brigade)
07-6-1091	Conduct a Gap Crossing (Battalion-Brigade)
07-6-1252	Conduct a Combined Arms Breach of an Obstacle (Battalion-Brigade)
17-6-1007	Conduct Intelligence, Surveillance, and Reconnaissance (ISR) Synchronization and Integration (Battalion-Brigade)
71-8-2321	Develop the Intelligence, Surveillance, and Reconnaissance Plan (Brigade-Corps)
17-6-3004	Employ Fires and Effects (Battalion-Brigade)
07-6-5037	Conduct Consolidation (Battalion-Brigade)
07-6-5082	Conduct Reorganization (Battalion-Brigade)
63-1-4032	Coordinate LOGPAC Operations

Table B-1. Example of Infantry battalion, IBCT METL (continued)

IN BN IBCT	MET (ART)
	TG (T&EO)
	Supporting Collective Task (T&EO)
71-8-5111	Conduct the Military Decisionmaking Process (Battalion-Corps)
71-8-5131	Execute Tactical Operations (Battalion-Corps)
71-8-5142	Evaluate Situation or Operation (Battalion-Corps)
03-2-9224	Conduct Operational Decontamination
07-6-1092	TG: Conduct an Attack (Battalion - Brigade)
07-6-1181	Conduct an Attack in an Urban Area (Battalion-Brigade)
07-6-1081	Conduct a Passage of Lines as Passing Unit (Battalion-Brigade)
07-6-6082	Conduct Mobility, Countermobility, and or Survivability (Battalion-Brigade)
07-6-1091	Conduct a Gap Crossing (Battalion-Brigade)
07-6-1252	Conduct a Combined Arms Breach of an Obstacle (Battalion-Brigade)
17-6-1007	Conduct Intelligence, Surveillance, and Reconnaissance (ISR) Synchronization and Integration (Battalion-Brigade)
71-8-2321	Develop the Intelligence, Surveillance, and Reconnaissance Plan (Brigade-Corps)
17-6-3004	Employ Fires and Effects (Battalion-Brigade)
07-6-5037	Conduct Consolidation (Battalion-Brigade)
07-6-5082	Conduct Reorganization (Battalion-Brigade)
63-1-4032	Coordinate LOGPAC Operations
71-8-5111	Conduct the Military Decisionmaking Process (Battalion-Corps)
71-8-5131	Execute Tactical Operations (Battalion-Corps)
71-8-5142	Evaluate Situation or Operation (Battalion-Corps)
03-2-9224	Conduct Operational Decontamination
ART 7.2	Conduct Defensive Operations
07-6-1028	TG: Conduct a Defense (Battalion - Brigade)

Table B-1. Example of Infantry battalion, IBCT METL (continued)

IN BN IBCT	MET (ART)
	TG (T&EO)
	Supporting Collective Task (T&EO)
07-6-1036	Conduct a Delay (Battalion - Brigade)
07-6-1144	Conduct a Withdrawal (Battalion - Brigade)
07-6-1082	Conduct a Passage of Lines as Stationary Unit (Battalion-Brigade)
07-6-6082	Conduct Mobility, Countermobility, and or Survivability (Battalion-Brigade)
17-6-1007	Conduct Intelligence, Surveillance, and Reconnaissance (ISR) Synchronization and Integration (Battalion-Brigade)
71-8-2321	Develop the Intelligence, Surveillance, and Reconnaissance Plan (Brigade-Corps)
17-6-3004	Employ Fires and Effects (Battalion-Brigade)
07-6-5037	Conduct Consolidation (Battalion-Brigade)
07-6-5082	Conduct Reorganization (Battalion-Brigade)
63-1-4032	Coordinate LOGPAC Operations
71-8-5111	Conduct the Military Decisionmaking Process (Battalion-Corps)
71-8-5131	Execute Tactical Operations (Battalion-Corps)
71-8-5142	Evaluate Situation or Operation (Battalion-Corps)
03-2-9224	Conduct Operational Decontamination
ART 6.7.3	Conduct Security Operations
17-6-9225	TG: Conduct a Screen (Battalion-Brigade)
07-6-1107	Conduct a Relief in Place (Battalion - Brigade)
17-6-1007	Conduct Intelligence, Surveillance, and Reconnaissance (ISR) Synchronization and Integration (Battalion-Brigade)
71-8-2321	Develop the Intelligence, Surveillance, and Reconnaissance Plan (Brigade Corps)
17-6-3004	Employ Fires and Effects (Battalion-Brigade)
63-1-4032	Coordinate LOGPAC Operations
71-8-5111	Conduct the Military Decisionmaking Process (Battalion-Corps)
71-8-5131	Execute Tactical Operations (Battalion-Corps)
71-8-5142	Evaluate Situation or Operation (Battalion-Corps)
17-6-9222	TG: Conduct a Guard (Battalion–Corps)

Table B-1. Example of Infantry battalion, IBCT METL (continued)

IN BN IBCT	MET (ART)
	TG (T&EO)
	Supporting Collective Task (T&EO)
ART 7.1.2	**Conduct Security Operations**
07-6-1082	Conduct a Passage of Lines as Stationary Unit (Battalion - Brigade)
17-6-3809	Conduct Battle Handover (Battalion-Brigade)
07-6-1107	Conduct a Relief in Place (Battalion-Brigade)
07-6-6082	Conduct Mobility, Countermobility, and or Survivability (Battalion-Brigade)
17-6-1007	Conduct Intelligence, Surveillance, and Reconnaissance (ISR) Synchronization and Integration (Battalion-Brigade)
71-8-2321	Develop the Intelligence, Surveillance, and Reconnaissance Plan (Brigade-Corps)
17-6-3004	Employ Fires and Effects (Battalion-Brigade)
63-1-4032	Coordinate LOGPAC Operations
71-8-5111	Conduct the Military Decisionmaking Process (Battalion-Corps)
71-8-5131	Execute Tactical Operations (Battalion-Corps)
71-8-5142	Evaluate Situation or Operation (Battalion-Corps)
71-8-5334	Plan Public Affairs Operations (Battalion-Corps)
ART 6.7.3	**Conduct Security Operations**
07-6-4000	Conduct a Civil Military Operation (Battalion-Brigade)
07-6-6073	Secure Civilians During Operations (Battalion-Brigade)
03-2-9224	Conduct Operational Decontamination
ART 7.3	**Conduct Stability Operations**
71-8-7331	**TG: Coordinate Essential Services for Host Nation (Brigade - Corps)**
17-6-9406	Conduct Lines of Communication Security (Battalion - Brigade)

Table B-1. Example of Infantry battalion, IBCT METL (continued)

IN BN IBCT	MET (ART)
	TG (T&EO)
	Supporting Collective Task (T&EO)
07-6-1272	Conduct Area Security (Battalion - Brigade)
17-6-1007	Conduct Intelligence, Surveillance, and Reconnaissance (ISR) Synchronization and Integration (Battalion-Brigade)
71-8-2321	Develop the Intelligence, Surveillance, and Reconnaissance Plan (Brigade Corps)
17-6-3004	Employ Fires and Effects (Battalion-Brigade)
63-1-4032	Coordinate LOGPAC Operations
71-8-5111	Conduct the Military Decisionmaking Process (Battalion-Corps)
71-8-5131	Execute Tactical Operations (Battalion-Corps)
71-8-5142	Evaluate Situation or Operation (Battalion-Corps)
71-8-5334	Plan Public Affairs Operations (Battalion-Corps)
07-6-4000	Conduct a Civil Military Operation (Battalion-Brigade)
07-6-6073	Secure Civilians During Operations (Battalion-Brigade)

This page intentionally left blank.

Appendix C

CATS Task Selection to METL Matrix

A CATS task selection to the company's METL matrix is an example that contains the list of CATS task selections specific to the weapons and antiarmor companies. Table C-1 contains specific CATS task selections that support the METs and task groups of the weapons and antiarmor companies.

The contents of the following table have been assembled from existing CATSs relating to the weapons and antiarmor companies, and is not complete. For more information regarding task selections relating to the reconnaissance troop refer to the CATSs found at ATN and DTMS.

Table C-1. Example of weapons and antiarmor company CATS task selection to METL matrix

Weapons and Antiarmor CO, Infantry Battalion IBCT		Unit		METs and Task Groups							
				Attack		Defend	Security			Stability	
Task Number	Task Title	Weapons Co	Antiarmor Co	Movement to Contact	Deliberate Attack	Area Defense	Screen	Recon	Area Security	Public Order & Safety	
06-TS-4339	Conduct Fire Planning and Prepare for Operations	X	X	X	X	X	X	X	X		
06-TS-4340	Conduct Occupation of the OP-FIST/COLT	X	X	X	X	X	X	X			

Table C-1. Example weapons and antiarmor company CATS task selection to METL matrix (continued)

Weapons and Antiarmor CO, Infantry Battalion IBCT		Unit		METs and Task Groups						
				Attack		Defend	Security			Stability
Task Number	Task Title	Weapons Co	Antiarmor Co	Movement to Contact	Deliberate Attack	Area Defense	Screen	Recon	Area Security	Public Order & Safety
06-TS-4341	Execute Fire Missions	X	X	X	X	X		X		
06-TS-4342	Conduct FIST/ COLT Team Operations	X	X	X	X	X		X		
07-TS-2472	Plan and Prepare for Operations (CO)		X	X	X	X		X	X	
07-TS-2477	Sustain Digital Proficiency	X	X	X	X	X		X	X	
07-TS-2871	Conduct Weapons/Anti Armor Company Operations	X	X	X	X	X		X	X	
07-TS-2872	Protect the Force (CO)	X	X	X	X	X		X	X	
07-TS-2873	Move Tactically (CO)	X	X	X	X	X		X	X	X
07-TS-2874	Conduct Offensive Operations (CO)	X	X	X	X	X		X		

Table C-1. Example weapons and antiarmor company CATS task selection to METL matrix (continued)

Weapons and Antiarmor CO, Infantry Battalion IBCT		Unit		METs and Task Groups							
				Attack		Defend	Security			Stability	
Task Number	Task Title	Weapons Co	Antiarmor Co	Movement to Contact	Deliberate Attack	Area Defense	Screen	Recon	Area Security	Public Order & Safety	
07-TS-2875	Defend (CO)	X	X			X	X	X			
07-TS-2876	Deploy/Redeploy the Company	X	X								
07-TS-2878	Sustain the Company	X	X								
07-TS-3150	TOW Gunnery Table XI (PLT Practice)		X								
07-TS-3151	TOW Gunnery Table XII (PLT Qualification)		X								
07-TS-3472	Plan and Prepare for Operations (PLT)		X	X	X	X	X	X			
07-TS-3871	Conduct Weapons/ Antiarmor Platoon Operations	X	X	X	X	X	X	X			
07-TS-3872	Protect the Force (PLT)	X	X	X	X	X	X	X			

Table C-1. Example weapons and antiarmor company CATS task selection to METL matrix (continued)

Weapons and Antiarmor CO, IBCT, SBCT		Unit		METs and Task Groups						
				Attack		Defend	Security			Stability
Task Number	Task Title	Weapons Co	Antiarmor Co	Movement to Contact	Deliberate Attack	Area Defense	Screen	Recon	Area Security	Public Order & Safety
07-TS-3873	Move Tactically (PLT)	X	X	X	X	X	X	X		
07-TS-3874	Overwatch/Support by Fire (PLT)	X	X	X	X	X	X	X		
07-TS-3875	Defend (PLT)	X	X				X	X	X	

Glossary

| Acronym/Term | Definition |

| --- | --- |
| 1SG | first sergeant |

A

AA	avenue of approach
ACE	ammunition, casualty, and equipment
AKO	Army knowledge online
AO	area of operation
APC	armored personnel carrier
ARFORGEN	Army forces generation
ASCOPE	areas, structures, capabilities, organizations, people, and events
ASV	armored security vehicle
ATGM	antitank guided missile
ATLDG	Army training and leader development guidance
ATN	Army Training Network
ATS	Army training strategy

B

BCIS	Battlefield Combat Identification System
BCT	brigade combat team
BCTC	battle command training center
BDA	battle damage assessment
BFSB	battlefield surveillance brigade
BHL	battle handover line
BN	battalion
BP	battle position
BRIDGEREP	bridge report

C

CALFEX	combined arms live fire exercise
CAS	close air support
CASEVAC	casualty evacuation
CATS	combined arms training strategy
CB	chemical and biological
CBRN	chemical, biological, radiological, nuclear
CCIR	commanders critical information requirement

Acronym/Term	Definition
CCP	casualty collection point
CCTT	close combat tactical trainer
CEF	contingency expeditionary force
CG	commanding general
COA	course of action
COIN	counterinsurgency
CoIST	company intelligence support team
COMMO	communication
COMSEC	communications security
COP	common operational picture
CP	command post
CPR	cardiopulmonary resuscitation
CSS	combat service support
CTC	combat training center

D

DE	directed energy
DEF	deployment expeditionary force
DOTD	Directorate of Training and Doctrine
DTMS	Digital Training Management System

E

EA	engagement area
ECOA	enemy courses of action
EOD	explosive ordnance disposal
EOF	escalation of force
EPLRS	Enhanced Position Location Reporting System
EPW	enemy prisoners of war

F

FBCB2	Force XXI Battle Command Brigade and Below
FIST	fire support team
FM	field manual
FMC	fully mission capable
FO	forward observer
FPF	final protective fires
FPL	final protective lines
FRAGO	fragmentary order
FSC	forward support company
FTX	field training exercise

Acronym/Term	Definition

G

GSR	ground surveillance radar

H

HBCT	heavy brigade combat team
HHC	headquarters and headquarters company
HN	host nation
HPT	high payoff target
HQ	headquarters
HQDA	Headquarters, Department of the Army
HUMINT	human intelligence

I

IBCT	Infantry brigade combat team
IMINT	imagery intelligence
IP	internet protocol
IPB	intelligence preparation of the battlefield
IR	intelligence requirement
ISR	intelligence, surveillance, and reconnaissance

J

JCATS	joint conflict and tactical simulation

K

KIA	killed in action

L

LD	line of departure
LDS	leader development strategy
LOA	limit of advance
LOGPAC	logistics package
LOGSTAT	logistics status
LRP	logistics release point
LVCG	live, virtual, constructive, and gaming
LZ	landing zone

M

MCoE	Maneuver Center of Excellence
MEDEVAC	medical evacuation
MET	mission-essential task
METL	mission-essential task list

Acronym/Term	*Definition*
METT-TC	mission, enemy, terrain and weather, troops and support available, time available and civil considerations
MILES	Multiple Integrated Laser Engagement System
MOPP4	mission-oriented protective posture 4
MP	military police
MTF	medical treatment facility
MTOE	modified table of organization and equipment
MTP	mission training plan
MWD	military working dog

N

NBC	nuclear, biological, and chemical
NCO	noncommissioned officer
NCOIC	noncommissioned officer in charge

O

OAKOC	observation, avenues of approach, key and decisive terrain, obstacles, and cover and concealment
OBSTINTEL	obstacle intelligence
OE	operational environment
OP	observation post
OPORD	operational order
OPSEC	operations security
ORP	objective rally point

P

PERSTAT	personnel status
PDDE	power-driven decontamination equipment
PIO	police intelligence operations
PIR	priority intelligence requirement
PMCS	preventive maintenance checks and balances
PME	professional military education
PMESII-PT	political, military, economic, social, information, infrastructure, physical environment and time
POL	petroleum, oil, and lubricant
POSNAV	position navigation
PSG	platoon sergeant

Q

QC	quality control

Acronym/Term	Definition

R

R&S	reconnaissance and surveillance
RA	regular Army
RC	Reserve Component
REDCON	readiness condition
REMBASS	Remotely Monitored Battlefield Sensor System
ROE	rules of engagement
ROI	rules of interaction
RP	release point

S

S-1	adjutant (Army)
S-4	supply officer (Army)
SALUTE	size, activity, location, unit, time, equipment
SBCT	Stryker brigade combat team
SCPE	simplified collective protective equipment
SIGINT	signal intelligence
SIR	specific information requirements
SITREP	situation report
SOEO	scheme of engineer operations
SOFA	status of forces agreement
SOI	signal operation instruction
SOO	support operations officer
SOP	standing operating procedure
SP	start point
SSI	signal supplemental instructions
STB	super tropical bleach
STT	sergeants time training
STX	situation training exercise
SU	situational understanding

T

T&EOS	training and evaluation outlines
TADSS	training aids, devices, simulators, and simulations
TAP	toxicological agent-protective
TC	training circular
TDA	table of distribution and allowances
TES	tactical engagement simulation
TG	task group
TLP	troop leading procedure
TOE	table of organization and equipment
TRADOC	Training and Doctrine Command
TRP	target reference point

Acronym/Term	Definition
TSOP	tactical standing operating procedure

U

UAS	Unmanned Aircraft System
UGS	unattended ground sensors
UTL	unit task list
UTM	unit training management

V

VBS2	Virtual Battlespace 2

W

WARNO	warning order
WFF	warfighting functions
WIA	wounded in action
WTPS	warfighter training support package

X

XO	executive officer

References

SOURCES USED

These are the sources quoted or paraphrased in this publication.

ARMY PUBLICATIONS

ADP 3-0, *Unified Land Operations*, 10 June 2011.

ADP 5-0, *The Operations Process*, 17 May 2012.

ADP 6-0, *Mission Command*, 17 May 2012.

AR 190-8, *Enemy Prisoners of War, Retained Personnel, Civilian Internees and Other Detainees*, 1 October 1997.

AR 350-1, *Army Training and Leader Development*, 18 December 2009.

AR 385-10, *The Army Safety Program*, 23 August 2007.

AR 600-8-1, *Army Casualty Program*, 30 April 2007.

ATTP 3-21.9, *SBCT Infantry Rifle Platoon and Squad*, 8 December 2010.

ATTP 4-02, *Army Health System*, 7 October 2011.

FM 1-02, *Operational Terms and Graphics*, 21 September 2004.

FM 2-01.3, *Intelligence Preparation of the Battlefield/Battlespace*, 15 October 2009.

FM 2-19.4, *Brigade Combat Team Intelligence Operations*, 25 November 2008.

FM 2-91.4, *Intelligence Support to Urban Operations*, 20 March 2008.

FM 3-07, *Stability Operations*, 6 October 2008.

FM 3-11, *Multi-Service Doctrine for Chemical, Biological, Radiological, and Nuclear Operations*, 1 July 2011.

FM 3-11.5, *Multiservice Tactics, Techniques, and Procedures for Chemical, Biological, Radiological, and Nuclear Decontamination*, 4 April 2006.

FM 3-19.4, *Military Police Leaders' Handbook*, 4 March 2002.

FM 3-20.15, *Tank Platoon*, 22 February 2007.

FM 3-20.96, *Reconnaissance and Cavalry Squadron*, 12 March 2010.

FM 3-20.971, *Reconnaissance and Cavalry Troop*, 4 August 2009.

FM 3-20.98, *Reconnaissance and Scout Platoon*, 3 August 2009.

FM 3-21.10, *The Infantry Rifle Company*, 27 July 2006.

FM 3-21.12, *The Infantry Weapons Company,* 1 July 2008.

FM 3-21.75, *The Warrior Ethos and Soldier Combat Skills*, 28 January 2008.

FM 3-21.8, *The Infantry Platoon and Squad*, 28 March 2007.

FM 3-21.91, *Tactical Employment of Antiarmor Platoons and Companies,* 26 November 2002

FM 3-28, *Civil Support Operations,* 20 August 2010.

FM 3-39, *Military Police Operations*, 16 February 2010.

FM 3-90, *Tactics*, 4 July 2001.

FM 3-90.1, *Tank and Mechanized Infantry Company Team*, 9 December 2002.

FM 4-0, *Sustainment*, 30 April 2009.

FM 4-02.2, *Medical Evacuation*, 8 May 2007.

FM 4-02.7, *Multiservice Tactics, Techniques, and Procedures for Health Service Support in a Chemical, Biological, Radiological, and Nuclear Environment*, 15 July 2009.

FM 4-25.11, *First Aid*, 23 December 2002.

FM 5-19, *Composite Risk Management*, 21 August 2006.

FM 6-22, *Army Leadership,* 12 October 2006.

FM 7-0, *Training Units and Developing Leaders for Full Spectrum Operations*, 23 February 2011.

FM 55-30, *Army Motor Transport Units and Operations*, 27 June 1997.

TC 3-34.489, *The Soldier and the Environment,* 8 May 2001.

The Army Leader Development Strategy for a 21st Century Army, <https://secureweb2.hqda.pentagon.mil/vdas_armyposturestatement/2010/information_papers/Army_Leader_Development_Strategy_for_a_21st_Century_Army_(ALDS).asp>

JOINT AND DEPARTMENT OF DEFENSE PUBLICATIONS

Most joint publications are available online at:
http://www.dtic.mil/doctrine/doctrine/doctrine_htm

JP 1-02, *Department of Defense Dictionary of Military and Associated Terms*, 8 November 2010.

OTHER PUBLICATIONS AND DOCUMENTS

None

DOCUMENTS NEEDED

FORMS

DA Form 1156, *Casualty Feeder Card.*

DA Form 2028, *Recommended Changes to Publications and Blank Forms.*

DD Form 1380, *U.S. Field Medical Card.*

READINGS RECOMMENDED

None

WEB SITES

Most Army doctrinal publications and regulations are available online at: http://www.apd.army.mil

Army, G-3/5/7 Memorandum, *A Leader Development Strategy for a 21st Century Army*, 25 November 2009. https://atn.army.mil/Media/docs/ArmyLdrDevStrategy_20091125[1].pdf

Army Knowledge Online (AKO) https://www.us.army mil

ATLDG, Army, G-3/5/7 Memorandum, *Army Training and Leader Development Guidance, FY 10-11*, 31 July 2009. https://www.us.army mil/suite/designer

Army Training Network (ATN) http://atn.army.mil/index.aspx

Combined Arms Training Strategy (CATS) https://atn.army.mil/dsp_CATSviewer01.aspx

Deputy Chief of Staff, G-3/5/7 Memorandum, *Army Training Strategy (ATS)*, 17 November 2009.
https://atn.army.mil/Media/docs/Army%20Training%20Strategy%20%20Appendix%20A%2017%20Dec%2009.pdf

Digital Training Management System (DTMS) https://dtms.army.mil (individual password required)

MCoE Collective Training Branch Home Page
https://www.us.army.mil/suite/grouppage/130823

Index

This page intentionally left blank.

By Order of the Secretary of the Army:

RAYMOND T. ODIERNO
General, United States Army
Chief of Staff

Official:

[signature: Joyce E. Morrow]

JOYCE E. MORROW
Administrative Assistant to the
Secretary of the Army
1135604

DISTRIBUTION:

Active Army, Army National Guard, and United States Army Reserve: To be distributed in accordance with the initial distribution number (IDN) 116020, requirements for TC 3-21.12.

www.ingramcontent.com/pod-product-compliance
Lightning Source LLC
Chambersburg PA
CBHW081946070426
42453CB00013BA/2273